Global Perspectives on Health Geography

Series editor

Valorie Crooks, Department of Geography, Simon Fraser University,
Burnaby, BC, Canada

Global Perspectives on Health Geography showcases cutting-edge health geography research that addresses pressing, contemporary aspects of the health-place interface. The bi-directional influence between health and place has been acknowledged for centuries, and understanding traditional and contemporary aspects of this connection is at the core of the discipline of health geography. Health geographers, for example, have: shown the complex ways in which places influence and directly impact our health; documented how and why we seek specific spaces to improve our wellbeing; and revealed how policies and practices across multiple scales affect health care delivery and receipt.

The series publishes a comprehensive portfolio of monographs and edited volumes that document the latest research in this important discipline. Proposals are accepted across a broad and ever-developing swath of topics as diverse as the discipline of health geography itself, including transnational health mobilities, experiential accounts of health and wellbeing, global-local health policies and practices, mHealth, environmental health (in)equity, theoretical approaches, and emerging spatial technologies as they relate to health and health services. Volumes in this series draw forth new methods, ways of thinking, and approaches to examining spatial and place-based aspects of health and health care across scales. They also weave together connections between health geography and other health and social science disciplines, and in doing so highlight the importance of spatial thinking.

Dr. Valorie Crooks (Simon Fraser University, crooks@sfu.ca) is the Series Editor of Global Perspectives on Health Geography. An author/editor questionnaire and book proposal form can be obtained from Publishing Editor Zachary Romano (zachary.romano@springer.com).

More information about this series at http://www.springer.com/series/15801

Gavin J. Andrews • Emma Rowland
Elizabeth Peter

Place and Professional Practice

The Geographies in Healthcare Work

Springer

Gavin J. Andrews
Department of Health, Aging and Society
McMaster University
Hamilton, ON, Canada

Emma Rowland
Florence Nightingale Faculty of Nursing
King's College London
London, UK

Elizabeth Peter
Lawrence S. Bloomberg Faculty of Nursing
University of Toronto
Toronto, ON, Canada

ISSN 2522-8005 ISSN 2522-8013 (electronic)
Global Perspectives on Health Geography
ISBN 978-3-030-64178-8 ISBN 978-3-030-64179-5 (eBook)
https://doi.org/10.1007/978-3-030-64179-5

The Cover image credit is: Gorodenkoff/shutterstock.

This Springer imprint is published by the registered company Springer Nature Switzerland AG
The registered company address is: Gewerbestrasse 11, 6330 Cham, Switzerland

Foreword

There is much to commend in this book. The first is bringing the discipline of geography, geographical topics and professional practice together. Covering a wide expanse of terrain, topics, issues and approaches, the team is well equipped to take forward an innovative agenda. Gavin Andrews—a health geographer who studies health care practice, Emma Rowland—a new scholar and nurse researcher with a PhD in human geography, and Elizabeth Peter—an established nurse ethicist with a career interest in ethics and place, all tackle geography and professional practice in different ways. By coincidence, I have had the pleasure of 'bumping' into each scholar in different venues, Gavin Andrews in the United Kingdom (UK), Elizabeth at the University of Toronto and Emma working in the same faculty as myself. We have discussed our intersecting interests in spatial dynamics of history, comparative geography of nursing workforce and emotional ecosystems of practice.

Space and place—the two core geographical concepts are of increasing importance in professional practice. At the time of writing, COVID-19 brings geography into stark reality with concepts such as social distancing, sheltering in place, transmission routes, care homes and the notion of 'community' as a setting all coming to the fore, not only as professional considerations but also in everyday public conversations. However, more generally, in the last thirty years, practice has been extended to a greater range of community places (such as homes, schools, the street) and has occurred in established settings (such as hospitals) that are ever-changing in their form and function. Moreover, technology has changed the spatialities and interactions between professionals and clients, particularly as remote care becomes more common via the telephone, internet and robotics.

Nurses talk often refers to the 'proximity' between themselves and their clients—their availability in space and time, and pressures on them. We talk of 'presence' and how, as physical and felt, it is important in therapeutic relationships and outcomes—where patients and clients come from, what routes they have taken to bring them to their current situations and how this shapes expectations, experiences and access to care. No where is this now more potent than geographical variations in the social determinants of care. Talk about 'therapeutic environments', both community and hospital-based, and what clients need to have 'in place' to maximise their experiences and outcomes is germane to how we organise care. As health professionals, we might not always think of all these things as geographical, but they clearly are.

But thinking geographically about care also helps us quite explicitly see how a geographical imagination and approach can be integrated into both research and practice. No previous publication has done this so far. Specifically, two of the middle chapters describe an applied qualitative geographical study in detail as an exemplar, and the final two chapters explicitly engage with well-known concepts and debates in practice and research, Skill Acquisition and Person-Centred Care, but add a new spin. There is also a new emergent vocabulary built into the methods and theoretical insights discussed in the book—'carescapes' and 'mobilities of care'. These chapters will be useful, not only to health professional researchers/academics thinking about geographical issues and frameworks but also to leaders in both management and professional practice and innovation. Students and clinical nurses will find much to relish and reflect upon in the reading. This book demonstrates that geography has arrived and can reframe and open new 'spaces' for analysis and the conduct of care.

Nursing Policy, Florence Nightingale
Faculty of Nursing and Midwifery
King's College London,
London, UK

Ann Marie Rafferty

Preface and Overview

This book is focused on geographical research on, and the geographical facets and dimensions to, health care work; on a broad range of work contexts, workplaces and work practices. It is perhaps the first book to consider these things explicitly and in-depth. *Conceptually,* the book conveys how space and place and related geographical ideas matter to clinical practice from the historical beginnings of care and medicine to the present day. *Theoretically,* it outlines the contributions of various traditions that have informed these understandings, ranging from classical 'in house' professional theories on the nature of clinical environments to ecology, spatial science, political economy, humanism, social constructionism and various post-structuralist and posthumanist theory. *Positionally,* the book showcases how a critical perspective can be usefully deployed at the very heart of medicine and health care research, exposing the many geographies in their (re)production. *Empirically,* it covers work across a range of job types (including physician, nurse, and multiple technical and therapeutic roles across multiple specialties). Finally, *informatively,* the book draws not only on the research of geographers but also on the research of those in other fields who have, over the years, conducted their own spatial/geographical studies, including in health services research, design and architecture, and the academic wings of various clinical specialities (most notably nursing research). Hence, it looks at the intimate geographies in practice (e.g. Andrews and Evans 2008) rather than macro-scale human resources, labour and workforce issues (e.g. Connell and Walton-Roberts 2016).

In terms of readership, we intend the book to be of interest to teachers and students; those leading or taking courses/modules focused on work and/or health care in human geography and other social science degree programs, and those leading or taking courses/modules in nursing and other health professional programs. Whilst, through the book, teachers might find fresh ways to convey the nature of health care work, students will find fresh perspectives that assist their understandings. We also intend the book to be of interest to researchers—indeed, to (post)graduate students and professional researchers, helping extend the scope of their research engagements with health care and health care work.

In terms of its academic position and rationale, the book comes at as distinct moment in time. As will become clear in the book, over a number of decades, both medical geography and the more recently emerging geography of health have displayed varied 'associations' with the topic of professional health care work. In medical geography, often using quantitative methods and with a conceptual emphasis on space, scholars have mapped important contexts to health care work (such as the distributive features of health and disease and the spatial accessibility and utilization of services), and even specific decisions made by workers (such as on where to work and where to refer clients); classic texts in this area include Joseph and Phillips (1984) and Meade and Earickson (2000). In the 'post-medical' geography of health, using qualitative methods and with a conceptual emphasis on place, scholars have focused largely 'downstream' from health care work on the consumption of health services (i.e. what one might think of as the net 'effect' of work), or otherwise have focused outside system interactions on experiences of illness and wellness (i.e. at times what might be the net 'neglect' of health care work) (Parr 2002, 2004). Meanwhile, most recently, changing things up considerably and filling the gaps left by above research, an increasing number of studies by geographers and by nurse scholars have begun to illustrate far more directly how health care roles and practices are shaped by space and place, and hence in research terms, how they might be understood through a geographical lens (see Andrews 2006, 2016; Andrews and Evans 2008). Engaging with and beyond this emerging literature, this book presents the first single comprehensive analysis that illustrates the vast breadth of geographical realities in health care work.

In terms of further justifying the book, it might be argued that all social science perspectives are relevant in approaching health care work because they reflect 'on-the-ground' practice realities that are important to understand. Sociology has proved insightful to professional practice because health care does not exist in a vacuum, and the social nature of it and its contexts are critical. Social psychology has proved insightful to professional practice because health care workers, their clients and the public ruminate and opinionate upon and make decisions on health matters. Economics has proved relevant to professional practice because workers increasingly take responsibility for, or are themselves, scarce resources (and so on). By the same token geography is insightful because, as suggested, all practice and purviews play out over space and place that determine their character (Andrews and Evans 2008). Indeed, showcasing the many on-the-ground geographical realities of clinical roles and practices, this book illustrates why geography is as important as any other discipline.

With regard to structure, Part I of the book is introductory and contextual. Chapter 1 examines how geographical ideas have been central to various types of practice, thinking and acting over many centuries and in many places. This includes labour divisions and forms of caring in prehistoric times, in the earliest origins of ancient Chinese and Indian (traditional) medicine, in the health ideas of ancient Greek writers, in medicine in the age of European exploration, in the nineteenth century origins of public health practice, in the work of Florence Nightingale and in mid-twentieth century nursing theory. The chapter explores how these historical strands have reached fruition in the current era with contemporary practice and research concepts such as 'work environments', 'clinical environments', 'environmental health', 'life-words' and broader movements such as environmentalism. Chapter 1 hence sets the scene for the ones that follow it, which are focused more on contemporary geographical thinking and ideas. Moving on, Chap. 2 reviews the contemporary geographical study of health care work. Initially, it examines key 'on-the-ground' transformations in health and health care which are fundamentally geographical in their making, form and consequences—hence, transformations which have demanded a geographical research perspective be taken as well-aligned vantage point with which to report and understand them and their impacts. These include increasing spatial diffusion of health professionals and roles; the transition of health care settings; the role of technologies in overcoming spatial limitations; the increasing emphasis on community and the social model of health; the embeddedness of geographical scales and concepts in policy and administration, and the globalization of work roles and responsibilities. It then moves on to examine how a number of academic developments—including in medical/health geography and other social and health sciences—have at the same time coalesced to provide an additional set of sectorial motivations and opportunities for geographical scholarship on health care work. Finally, it describes the main theoretical traditions and approaches which have constituted 'geographies of health care work' as a multidisciplinary academic enterprise and field—including spatial science, political economy, social constructionism/humanistic and most recently non-representational—providing examples of recent empirical research which has been framed by each.

Part II of the book is exemplary, Chaps. 3–5 being empirical case studies based on two of the authors' original empirical work, that showcase contemporary geographies in professional practice. Chapter 3, the first of these, presents data derived from ethnographic observation of four hospital wards and interviews with individuals working in them (together they representing multiple work roles/categories). The aim is to articulate the character and function of 'emotional geographies' in hospital-based/ward work. Specific themes include geographies of emotional detachment, contested emotion, emotional attachment, emotional containment and communicating emotions. Chapter 4, the second of the case studies, presents data derived from ethnographic observations of ambulance work and interviews with ambulance crew, the aim being to describe the nature and impact of particular spaces on this form of work including ambulance stations, mobile workspaces, public spaces and private spaces. Similar emotional categories to Chap. 3 are explored. Moreover, across many of these spaces, a number of experiences are mapped, including frustrations, stresses, coping strategies, gendered dimensions and bodily agency (body categorization, practices, transformations, routines and rituals). Indeed, this chapter showcases issues related to professional health care work across non-traditional community spaces. Chapter 5, the final case study, presents an analysis of published policy, institutional, legal and media statements relating a case where a nurse was found to be extremely 'dangerous'. The idea is to think about how trust in ethical nursing work, and the places of nursing work, is represented. Attention is paid to the involvement of different scales in work and its regulation and its representation (international, national, city and setting), and how each is used/implicated.

Part III of the book is more visionary. The final two chapters showcase the potential for a more integrated geographical contribution to professional practice. How geography might be elevated beyond its current role as a 'useful' research lens and perspective by giving the discipline a more central voice in mainstream health professional debates. To this end, Chap. 6 showcases how geography might play a fuller and more embedded role in understandings of 'person-centred care' (PCC). After reviewing traditional research areas on PCC, it describes how PCC might be understood as a 'more-than–human' spatial production. Specifically, it describes how a posthumanist theoretical orientation might rethink some concepts central in PCC (identity and meaning, emotions, difference and oppression, agency and communication). Then, using a recently developed three-part posthumanist typology, it considers how PCC might emerge and express within material social assemblages; be enacted and performed affectively by vital bodies and vibrant objects; and be conducted in immediate, pre-personal, more-than-representational spacetimes. Chapter 7 showcases how geography might play a fuller and more embedded role in understandings and debates on 'skills'. Specifically, after briefly reviewing traditional approaches to studying and understanding skill in health professional research, it takes principles of non-representational theory, and proposes some ways forward for future research through a deeper ontological understanding of skill and its processual, spatial emergence—summarised by us in the term 'skilling space'. Finally, at the end of this last chapter a nod is given to the broader research agenda as we move forward with geographical perspectives.

Hamilton, ON, Canada — Gavin J. Andrews
London, UK — Emma Rowland
Toronto, ON, Canada — Elizabeth Peter

References

Andrews, G. J. (2006). Geographies of health in nursing. *Health & Place, 12*(1), 110.

Andrews, G. J. (2016). Geographical thinking in nursing inquiry, part one: Locations, contents, meanings. *Nursing Philosophy, 17*(4), 262–281.

Andrews, G. J., & Evans, J. (2008). Understanding the reproduction of health care: Towards geographies in health care work. *Progress in Human Geography, 32*(6), 759–780.

Connell, J., & Walton-Roberts, M. (2016). What about the workers? The missing geographies of health care. *Progress in Human Geography, 40*(2), 158–176.

Joseph, A. E., & Phillips, D. R. (1984). *Accessibility and utilization: Geographical perspectives on health care delivery*. Thousand Oaks: Sage.

Meade, M. S., & Earickson, R. (2000). *Medical geography*. New York: *Guildford Press*.

Parr, H. (2002). Medical geography: Diagnosing the body in medical and health geography, 1999–2000. *Progress in Human Geography, 26*(2), 240–251.

Parr, H. (2004). Medical geography: Critical medical and health geography? *Progress in Human Geography, 28*(2), 246–257.

Contents

About the Authors

Gavin J. Andrews, BA, PhD is a professor at the Department of Health, Aging and Society, McMaster University, Canada (and an associate member of the Department of Geography and Earth Sciences at the same institution). As a health geographer, his wide-ranging interests include the dynamics between space/place and: health care education and work, nursing, holistic medicine, aging and fitness cultures. Much of his work is positional and considers the development, state-of-the-art and future of health geography. In recent years, he has become interested in the potential of posthumanism and non-representational theory in conveying the vital 'taking place' of health and health care. Department of Health, Aging and Society, KTH 240, McMaster University, Hamilton, ON, Canada

Emma Rowland, BA, PhD is a lecturer at the Florence Nightingale Faculty of Nursing and Midwifery, King's College London, United Kingdom. She is a geographer with an interest in both emotional and health geographies. Her work focuses on how space, place, temporality and ideas of proximity and distance within secondary care settings (hospital and ambulance service), impact on health professionals' emotion management, relationships with each other and their patients and on their delivery of patient care. Her current scholarly activities focus on the emotion management, specifically emotion work of partners with a spouse affected by hereditary breast cancer. The Florence Nightingale Faculty of Nursing, Midwifery and Palliative Care, King's College London, London, UK

Elizabeth Peter, RN, PhD is a professor at the Lawrence S. Bloomberg Faculty of Nursing and a member of the Joint Centre for Bioethics and the Centre for Critical Qualitative Health Research, University of Toronto, Canada. Her scholarship reflects her interdisciplinary background in nursing, philosophy and bioethics. Drawing on the work of human geographers, she has used geographical concepts to examine the unique ethical concerns that arise in the delivery of home and community services. She has also explored the spatial dimensions of moral distress in nursing that arise as a result of nurses' proximity to patients. Lawrence S. Bloomberg Faculty of Nursing, University of Toronto, Toronto, ON, Canada

Part I
Introductions

Chapter 1
The Geographical Origins of Geographical Thinking on Health Care Work

Gavin J. Andrews, Emma Rowland, and Elizabeth Peter

Abstract This chapter explores how geographical ideas have been central to thinking and acting in caring practice over many centuries in many places. This includes in forms of caring in prehistoric times; in the earliest origins of ancient Chinese and Indian medicine; in the health ideas of ancient Greek writers; in medicine in the age of European exploration; in the nineteenth century origins of public health practice; in the work of Florence Nightingale, and in mid-twentieth century nursing theory. The chapter then explores how these historical strands have reached fruition in the current era with contemporary practice and in research concepts such as 'work environment', 'clinical environment', 'environmental health', 'lifeworlds' and in broader movements such as environmentalism. The chapter sets the scene historically for the ones that follow it in the book.

Introduction

This opening chapter is unique in that it looks back in time in order to establish some foundations and precedent for the book, considering the origins of geographical thinking on health care work and practice. Telling this history is challenging because different versions of it exist. As Valencius (2000) notes, there is certainly no one universally accepted story of geographical thinking in this area. Current scholars with different backgrounds—be they geographers, historians, medical historians or physicians—have written quite different accounts that emphasize the work of particular ancestors that tend to support their own disciplinary allegiances and priorities. Nevertheless, we think we can do a more wide-ranging job than others have done. Structured chronologically, the chapter moves quickly between diverse time periods conveying how, prior to contemporary scholarship, geographical thinking was undertaken predominantly by clinicians with a view to improving the health states of the individuals and populations under their particular spheres of responsibility and improving the nature of their practice directly. Hence there are two take home messages. First, that the origins of geographical thinking on health care work and practice have reflected their particular geographical contexts and situations.

G. J. Andrews et al., *Place and Professional Practice*, Global Perspectives on Health Geography, https://doi.org/10.1007/978-3-030-64179-5_1

Second, and perhaps most fundamentally, because these people were some of the earliest thinkers and actors of any on health care, they and their geographical ideas have been central to developments and priorities in the overall sector over many centuries, as we shall see from the very earliest forms of interpersonal care and medicine, to later tropical and social medicine, through to more recent eras and the idea of 'practice environments'.

The Prehistorical Origins of Place-Based Care in Archaic Human and Modern Human Populations (450,000 BC+)

We start the story from the very beginnings of humankind, not typical in academic reviews. Recent anthropological research has identified forms of what would now be categorized as care in the earliest archaic human populations. Forms of care that were very much a product and reaction to the environments that these populations had to live and survive in (Spikins et al. 2018a, b). Laying the basis for care was the development of compassion in populations. In terms of definitions, psycho-anthropological research tells us that to show compassion is to show sympathy and concern for various forms of suffering that others might experience, this resulting in forms of tolerance and sensitivity, both of which lay the foundation for kind and supportive acts. Indeed, as Spikins et al. (2010) note, compassion is feeling emotion for another individual's emotion. It is free from immediate obligation, often being spontaneous and not thought out, yet as an 'idea' might be incorporated into rational thoughts and planning (Spikins 2015; Spikins et al. 2010).

Covering a broad swathe of prehistoric history, Spikins et al. (2010) note four stages or thresholds of compassion existing in archaic humans and modern humans (Homo Sapiens). Each differs from thresholds of compassion observable in more primitive primate groups (such as chimps and other great apes), which, although capable of displaying concern and some forms of care, usually only extend this to young and/or for limited periods: (1) fleeting responses to distress (6.2 million years ago); (2) regulated emotion and rational thought (1.8 million years ago); (3) deep seated commitment to welfare of others, including long term planning (300,000–50,000 BC); (4) moral obligation, including the development of abstract concepts—such as shown in symbolic material objects and art about the duty to care (100,000 BC). As Spikins (2017) explains, however, it was Neanderthal populations (archaics of 450,000–40,000 BC) that displayed most progress in terms of compassion, evidence suggesting that family and social groups would look after vulnerable members for as long as was required, often for their full lives (evidence existing, for example, of feeding, sharing food and water, and protection). These kinds of concerns and acts, far from being wasted energy, constituted the basis for collaboration and species development, and certainly challenge the popular stereotype of Neanderthals being mindless thugs engaged purely in a game of survival of the fittest (Spikins 2017). As Spikins suggests, there is a practical 'logic' to compassion and its utility, it being passed on through natural selection. The most successful individuals and groups would display greater degrees of compassion (hence displaying one's emotional credentials attracted a mate). Through compassion, then, what eventually developed is complex emotional minds in early humans. These in turn allowed even greater degrees of trust, networking, role definition and perseverance in technical skills to emerge, and overall more rapid social development (Spikins 2017).

What was critical to species development in Neanderthals was the specific environments they were attempting to survive and prosper in, the activities they had to undertake for this to happen, and the specific instances where compassionate acts were most useful and necessary. As Spikins et al. (2018a, b) explain, Neanderthals lived in the cool temperate zone climates of what is now Europe and Northern Africa, which were prone to great seasonal temperature variations and periods of glaciation. They existed here because of the abundance of large mammals—megafauna—which were their primary source of food. However, hunting these mammals involved quick movement over terrain and close range killing, both of which frequently resulted in injuries (evidence of injury resulting from inter-group/human aggression and conflict being less clear (Spikins et al. 2018b)). Research suggests that in order to be more successful Neanderthals gradually became more knowledgeable in approaches to injury management. Indeed, the archeological record shows that their 'primary care' included applying stints to breaks, cleaning and covering wounds, administering water for fever management, and eventually administering types of acids as early painkillers and antiseptics. Meanwhile, 'after care' for the sick included providing shelter, protection and food sharing. This often lasted weeks or however long it took for injuries to healing and infections to clear (Spikins et al. 2018a, b). Indeed, research describes how most older Neanderthal skeletons show at least one injury, suggesting that such care was common and effective (Spikins et al. 2018b). Meanwhile, research argues that labour reallocation was also undertaken for mutual benefit; injured or recovering group members being given less physical tasks such as childcare, tool or clothing manufacturing, cooking or fire tending (Spikins et al. 2018a, b). What is not known though is how primary or after care might have been allocated between group members, although it is logical to assume—given distributed behaviors and roles known to exist in numerous other primate groups—that certain group members might have been more expert and involved in caring than others (Spikins et al. 2018a, b). If so, this would constitute the very earliest form of specialization in care.

In sum, in terms of overall contribution, in Neanderthal populations caring for injured members of a group constituted a form of 'risk pooling'. Knowing that if injured you would be cared for, allowed individuals to take greater risks whilst hunting and subsequently to achieve greater success for the benefit of the group (Spikins et al. 2018a, b). Moreover, as part of this, caring practices showed a basic understanding of the risks of non-care and a basic understanding, if not of disease pathology and physiology, at least of the consequences of disease pathology and physiology (Spikins et al. 2018b). The extent to which caring in Neanderthal populations constitutes conscious geographical 'thinking' on health care practice is of course debatable. Most likely, it is a form of practice almost automatically responsive to particular geographical circumstances. Yet, it certainly represents the start of a relationship between geography and health that would continue for half a million years. Indeed, it would continue through the remainder of the prehistoric period in the increasingly sophisticated deployment of herbs, clays, basic dentistry, hygiene and wound management by modern humans during the Middle Stone Age (250,000–50,000 BC), New Stone Age (50,000–3500 BC), Bronze Age (3500–1000 BC) and Iron Age (1000–500 BC) (Arnott 1996), eventually coming to full fruition in the historical periods examined next.

Medicine's Origins in Climate and Nature: From Ancient Greece to the Ancient East (Fifth Century BC to the Fifth Century AD)

Across historical reviews of medicine and geography, it is frequently argued that the origins of the coming together of these two traditions lies in ancient Greece and the '*Hippocratic Corpus*'; albeit, as we shall see, this is a claim that some have challenged (Barrett 2000a). Much has been said about the Corpus, which is constituted of over 60 individual works written over a 70-year period. To summarize, it presented new thinking on medicine and an approach which for the first time in human history: (1) was committed to understanding disease and care through systematic deductive thinking; (2) recognized disease as a physical and natural phenomenon (not a result, for example, of divine intervention or magic, hence subject to same laws as the rest of the universe); (3) rejected monoism (single cause explanations for disease), recognizing a plurality of causes and effects; and (4) possessed a commitment to diagnostic accuracy yet was written simply for a broad intellectual and practitioner audience. The specific work in the corpus of most interest in terms of geographical thinking is *Airs, Waters and Places* (400 BC). Indeed, as Barrett (2000a) notes, this document provided guidance for the travelling physician of the time on what to expect when entering a new location. It conveyed in simple terms that disease was a product of local environmental factors, particularly water sources, the direction of prevailing winds and seasonal climatic variations. Mirroring the known world at the time, Hippocrates' observations were made in the context of the peoples and landscapes of Western Asia and Northern Africa, although they were not place-specific. As an example of how *Airs, Waters and Places* is represented as a building block, and the attention it still receives to this day, the fall 2017 edition of Harvard University's Public Health Magazine reads:

> Climate's profound influence on health is not a new story. Nearly 2,500 years ago, in a treatise titled *On Airs, Waters, and Places*, Hippocrates advised traveling physicians to "consider the seasons of the year, and what effects each of them produces." He went on to expound upon the health implications of "the winds, the hot and the cold," "the qualities of the waters," rain and drought, each city's unique setting in the landscape, and even whether

its inhabitants were given to excess and passivity or discipline and courage. "These things one ought to consider most attentively," he wrote (Harvard 2017)

The importance of *Airs Waters and Places* is not just a contemporary realization. As Valencius (2000) notes Roman scholars and healers, such as Galen, commented on how the document emphasized physical properties (hot, cold, wet, dry) and related causal elements (earth, air, water, fire) in explaining health. This recognition later led, in the first century AD, to other Roman scholars and clinicians providing powerful descriptions of the urban to rural spread of contagious diseases and their impacts in population centres (Barrett 2000a). Later in the centuries that followed, *Airs Waters and Places* was discussed explicitly in the theorizations of many scholars and traditions (covered in the following sections in this chapter).

As mentioned earlier, some have questioned that *Airs Waters and Places* constitutes the lone, and very earliest example of medicine and geography coming together from which all else is proceeds (Barrett 2000a). Barrett challenges conventional wisdom, tracing the origins of geographical thinking on medicine to other times and places, providing a more complex and richer picture. In particular, Barrett notes how seminal works in ancient Chinese medicine (Huang Ti Nei Ching Su Wen) and Indian medicine (Ayurveda) from the same period as *Airs Waters and Places* connected environment, diet and lifestyle with the occurrence of specific diseases. With regard to the former, from the late centuries BC, Su Wen (to use the common abbreviation) helped develop early thinking on vessel theory and *Yin-yang qi* established fundamental practices such as acupuncture that continue to the current day (Unschuld 2003). Also notable from a geographical perspective are the climatological theories of *Wu yun liu qi*, particularly its description of how weather (due to the changing cosmological forces between earth and space), broader environmental conditions, climatic agents and human behavior determine the emergence of disease. With regard to Indian medicine, the geographical orientation in Ayurveda is all about the need for humanity to live in tune with nature. Indeed, Ayurveda is based on the idea that being part of nature, human life and behaviour should not conflict with it. The argument being that if humans are in harmony with nature, human desires will inevitably be only for what nurtures health (conversely if humans are out of tune with nature, human desires will be for things that cause ill-health). Moreover, it is believed that synchronisation of human and nature rhythms is important for obtaining uplifting sensory experiences of nature in the day (tastes, touches, sounds, smells, sights) and for obtaining restful sleep at night.

Faith, Herbalism and Humorism in the Middle Ages/Medieval Period (the Fifth to the Fourteenth Century AD)

A number of trends that emerged in Europe and the Middle East during the middle ages helped evolve and shape geographical ideas in medicine, some of which were to do with new belief systems, and some of which were to do with the emerging structure of society at the time. With regard to belief systems, the rise of Christianity during the 'age of faith' linked nature, landscape and religion under explanations for health (Barrett 2000a). Meanwhile scientific curiosity for knowledge emerged at this time, the priority and 'need to know' more about health in places arising as an ethical imperative, part of an early pre-renaissance, pre-scientific curiosity for knowledge (Barrett 2000a). With regard to the structure of society, both higher concentrations and greater mobilities of population occurred in this period and, with these changing dynamics, greater incidence of diseases. As Barrett observes, in Europe the era of the 'great pandemics' (such as the Black Death) ushered in explanations on causes of diseases that, for the first time, attempted to account for where humans are located and move.

At this time, two related systems and principles of medicine emerged: herbalism and humorism. Herbalism originated in ideas developed in Greek and Roman texts that had survived to the time. It continued in Pagan folk medicine (from the second century AD), and eventually developed into a fully-fledged discipline by the middle ages. Key collections of texts, such as *Pseudo-Apuleius*, included illustrations of common and familiar plants that could assist health. Herbal healing was heavily formed and structured by religion, it based on the idea that although god might punish sin and immorality with illness; god made nature that can cure disease. Indeed, as Christianity became involved in herbalism, monasteries emerged as centers for herbal healing, facilitating simultaneous possibilities for both prayer and treatment. They had workers—such as monks with specialist knowledge on cultivation—to organize both thought and practice, and also extensive gardens for the growing of relevant plants. Indeed, the first books on herbal remedies were produced by monasteries, notably the Welsh *Red Book of Hergest* at the start of the fifteenth century (Barrett 2000a).

Humorism, meanwhile, underpinned the practice of herbal healing. As a belief system it observed four humours or principal fluids—black bile, yellow bile, phlegm and blood—each produced by particular organs in the body. The idea was that these have to be in balance for a person to be in good health (diet, bloodletting, coughing leeches being used in practice to restore balance). Notably the four humours were aligned with the four seasons and their nature experiences: black bile to autumn, cold/dry and earth; yellow bile to summer, warm/dry and fire; phlegm to winter, cold/wet and water, and blood to spring, warm/wet and air. Key figures influenced developments at the time, most notably Hildegard of Bingen 1098–1179 (Saint Hildegard), a Hippocratic informed practitioner who developed her own herbal remedies and medicinal texts (the most influential of the latter being *Causae et curae*). Both herbal healing and humorism were sustained in Europe by training and apprenticeships formally organized by the Catholic church, this being an early forerunner to modern institutionalized mass medical education (Barrett 2000a).

Geographical Analysis and Early Cartography in Scientific Social Medicine (Seventeenth to Nineteenth Centuries)

Informed by the advent of modern scientific medicine—and its founding disciplines such as biomedical research, microbiology and pharmacology—and necessitated by increasing urbanization and poor urban conditions, by the seventeenth century the idea of social medicine started to emerge; its interest being in how social circumstances and phenomenon can impact upon health. The history of early social medicine and public health more broadly is detailed and well-trodden. It is clear, however, that geographical strands of thinking and scholarship run strong throughout it, emanating from different times and places. As Valencius (2000) notes, these strands date as far back as Thomas Sydenham's work on London epidemics, 1661–1675, that for the first time forged a formal quantitative research approach to studying urban environments and disease. Later, as Barrett (2002) notes, they emerged in the work of French physicians, and entries they made in medical dictionaries and encyclopedias of the time including Jean-Noël Hallé (1787, 1792) in *Encyclopédie Méthodique*, and Julien Virey (1817) in *Dictionnaire des Sciences Médicales*. Otherwise geographical strands of thinking and scholarship emerged in the pioneering work of Louis-René Villerme who studied mortality rates in the districts of Paris and the health of cotton workers in different parts of France, the latter published in 1840 (notably in 1829 he founding the first academic journal of urban studies on health *Annales d'hygiène publique et de médecine légale*). Moreover, as Barrett (2000b) suggests, these strands emerged across the Atlantic in the work of New York physician Valentine Seaman (1798), which included a "spot map" of the lower East Side street locations of occurrences of Yellow Fever.

These origins acknowledged, the nineteenth century geographical work of three physicians and a clinician/nurse is particularly notable for its contribution to the social medicine, much of it arising in the same decade. The first of these physicians is John Snow. As McLeod (2000) describes, his 1854 map showed clustering of cholera cases around a Broad Street (London) water pump. The key advancement in this research not being the mapping of disease—which had been done before—but the fact that it established cause through mapping (albeit that this history has been challenged (McLeod 2000)). The second physician is Daniel Drake. As Barrett (1996) describes, an American scholar, Drake published two volumes titled *Principle diseases of the interior of North America* (Drake 1850, 1854), which were manuals for physicians that emphasized how hydrographic regions, watersheds, local geology and climate and ethnic groups relate to different categories of illness. Whilst volume one included numerous micro-observations (such as types of soil or water flow or quality) and observed basic spatial patterns of disease, volume two was focused a little more on symptoms and treatment. The third physician is Alfred Haviland. As Barrett (1998) observes, this English scholar used national mortality statistics to predict causes of heart disease, cancer and tuberculosis (attributing prevalence to air quality and movement, soil type and drainage), his three key books being '*Climate, weather and disease*' (Haviland 1855); '*The geographical distribution of heart disease and dropsy, cancer in females and Phthisis in females in England and Wales'* (Haviland 1855); and *'The geographical distribution of disease in great Britain'* (Haviland 1892). The fourth, and perhaps the most well- known and influential, is Florence Nightingale. As Andrews (2003) describes, the founder of modern nursing's key work *Notes on Nursing: What it is and What it is Not* (Nightingale 1859) had both a micro scale focus on practice settings (as is discussed later in this chapter) and a wider macro-scale emphasis on social medicine, together they laying a heavy geographical cornerstone in the very earliest foundations of modern nursing. Nursing, according to Nightingale, was about more than caring tasks, incorporating a fundamental knowledge of, and some degree of responsibility for, the places where patients and potential patients find themselves (Andrews 2003, 2016). Indeed, Nightingale considered living and sanitary conditions in cities, and briefly outlined the scope and extent of health problems facing British society. Her description of the lack of urban cleanliness painted a horrifying picture of London life with 'dung heaps' in the streets contributing to child epidemics and illness.

In sum, historians of medicine often regard geographical scholarship as being central to the development of social medicine particularly from in the early nineteenth century onwards. In this way geography has been a political tool; a way for physicians and later nurses to assert their ideas and professions, and a way to help them establish new intellectual and physical territories (Valencius 2000).

Tropical, Colonial Geographical Medicine: A Focus on 'Exotic' Places (Eighteenth and Nineteenth Centuries)

Precedent for what is known as 'geographical medicine' exists in the late middle ages and the earliest colonial exploration and expansion of this period. In the fourteenth century during European 'discovery' of new lands and peoples, scholars began to consider the climatic and other conditions that led to disease and death, often describing the afflictions affecting explorers themselves through the use of detailed record keeping on the health of ships' crews and shore-based teams (Barrett 2000a). This early attention laid the basis for the later work of key scholars, one of these being James Lind, a Scottish physician and pioneer of naval medicine. Lind studied diverse phenomenon ranging from ship and onshore diets and their impact on sickness, to water quality, humidity and soil type as causal factors for fever. As Barrett (1991) suggests, Lind's 1768 book, *'An Essay on the Incidence of Diseases in Hot Climates'*, is of great significance based on his research and knowledge gained practicing around the world. Indeed, the book has three parts. The first part divides the world into the four disease regions; Europe and North America, Africa, the East Indies and the West Indies. The second part gives advice on the preservation of health for those who live in coastal areas in hot climates, and those who live inland (focusing on the role of site, location, and the characteristics of place). The third suggests cures for the particular illnesses. Barrett (1991, p. 348) provides a passage that illustrates Lind's critical reasoning:

> Men who thus exchange their native for a distant climate, may be considered as affected in a manner somewhat analogous to that of plants, removed into a foreign soil; where the utmost care and attention are required, to keep them in health, and inure them to their new situation; since, thus transplanted some change and alteration must happen in the constitutions of both. Some climates are healthy and salutary to European constitutions; as some soils are favourable to the production of European plants. But the countries beyond the limits of Europe which are chiefly frequented by Europeans, are very unhealthy, and the climate often proves fatal to them (pp. 2–3).

Another key scholar was August Hirsch, a physician and medical historian who wrote on cholera, plague and malaria in specific regional contexts. As Barrett (2000b) notes, through works such as his three volume *Handbuch der historisch-geographischen Pathologie*, 1859–1864, Hirsch's critiqued geographical medicine of the period and proposed rigorous scientific principles in future studies (such as comparative analysis between groups and places). Hence Hirsch established approaches that would see a new era of geographical medicine more widely accepted as a bonified medical science (Barrett 2000b). A final key scholar was L.L. Finke, a German physician who helped develop early medical cartography at the time (an important technical accompaniment to the type of substantive studies noted above). Finke likely produced the first world map of disease in the late eighteenth or early nineteenth century (Barrett 2000c). As Valencius (2000) suggests, through the works of Lind Hirsch, Finke and others, progress in geographical medicine was substantial. The improvements they made in measurement, statistical methods and mapping—along with the rise of state and other organizations dedicated to environment and health—meant that the discipline became an early legitimate 'big science'.

It is important, however, to take a critical post-colonial perspective given the roots of geographical medicine in empire building. As Valencius (2000) notes, geographical medicine was very much a tool of imperial power, directed from the centres of global power and exerted through their outreached powers. As a discipline it helped understand, change and improve local health conditions, yet always as a 'favor' to local populations, and often tackling diseases spread by imperialism in the first place. Moreover, it practically and directly served imperialist missions through its objective to improve the health of occupying military and administrative forces, hence improve their efficiency and their ability to maintain social control (see Anderson 1998; Crozier 2007; Ernst 2007; Marks 1997). Ultimately, then geographical medicine colluded in empire building as a technical discourse (Anderson 1998).

From Bacteriology to the Emergence of Medical and Health Geography (Nineteenth and Twentieth Centuries)

The rise of bacteriology in the late nineteenth century played an important role in geographical thinking on medicine. A few early geographers had been studying germs sporadically and in an rudimentary fashion prior to this period, but bacteriology quickly took over as the dominant approach establishing a new medical science with a spatial focus (Mitman and Numbers 2003; Numbers 2000). Indeed, new discoveries in this period changed the practice of medicine (notably in the work of Pasteur, Kotch and others on germ theory and eventually disinfection and vaccination). As Barrett (2000a) notes, it was an outwards expanding focus that was a key development. Scholars at the time needed to see how their ideas worked out outside the laboratory, and how they might address pressing population health problems. This led to physicians and others studying the relationships between diseases and geographical phenomenon that might cultivate them or assist vectors of transmission (e.g. temperature, humidity, elevation, soil composition, pollution). In this vein, key books of the era include *Disease pathology* (Davidson 1892); *The geography of disease* (Clemow 1903); *Geographie Medicale* (Laurent 1905) and, as noted earlier, *The geographical distribution of diseases in Great Britain* (Haviland 1892).

Although interest in bacteria continued in the early twentieth century and during the interwar period, new interests also emerged, particularly on state and health systems. As Barrett (2000a) describes, a geopolitical focus developed that included an emphasis on the role of governments in national health. Research in Germany, for example, compared and contrasted socialist, capitalist and fascist approaches. Specific developments and ideas at the time included 'geographical racial pathology'—an environmentally determinist approach that attributed disease in certain races to certain places. This work, like other geography at the time, has racist motivations or implications and conflict messages (for example, work by Zeiss that discussed unhealthy eastern Europeans invading and infecting healthy Germans, and work by Ratzel that promoted the idea of Lebensraum as a weapon/barrier to protect biological purity). Although this was a fleeting and rather unfortunate research interest, political interests were revisited later in far more politically correct and less ideological forms in contemporary medical and health geography that takes a political economy perspective.

By the mid twentieth century, medical geography emerged as the discipline we recognise today. As Barrett (2000a) suggests, part of this development involved far more social scientists studying disease geographically. This reduced the proportion of pure clinicians involved in scholarship although undeniably some medical geographers were still medically trained, and medical geography was still broadly 'servicing' medicine's needs and agendas (as well as attaching itself to medicine for credibility purposes and legitimacy). A key figure of the time Jacques May, a colonial surgeon of tropical medicine, established early methods and objectives of medical geography (May 1950) and introduced and ecological approaches (May 1959). The latter described the interplay of pathogens and geographical factors as 'geogens', this being the discipline's first 'in-house' theoretical approach, which separated it from geographical medicine (Brown and Moon 2004). As Brown and Moon note, May's work was multifaceted, particularly as it interweaved geographical medicine and medical geography and allowed them to inform each other (although still being firmly convinced in the hegemony of scientific medicine and its ability to simultaneously save developing world populations and improve military efficiency).

In the latter half of the twentieth century, medical geography expanded to broadly include two streams of study both relational somewhat to medicine; the geography of disease (i.e. the concern of medicine) and the geography of health care (i.e. the provision of medicine). More latterly, the 'geography of health' emerging as a third stream focused, often qualitatively, on subjects including health status experience and attainment, wellbeing and public health (for associated typologies and debates see Kearns 1993; Kearns and Moon 2002; Meade and Earickson 2000; Paul 1985). As Phillips (1985) suggests, the latter two streams signified a break from past disciplinary building blocks, as theoretically they owed as much to developments in economic, social, urban and cultural geography as they did to medicine. Theoretically, they have included positivistic, political economy, humanistic, social constructivist and, most recently, poststructuralist and posthumanist/non-representational theoretical approaches. These will be examined in the remaining chapters of this book.

Nightingale and the Origins of 'Nursing Environment' (Nineteenth Century)

Whilst most of the aforementioned historical traditions are about physicians being aware of their macro-scale socio, bio and natural contexts and how they affect their practice, in Florence Nightingale's work we see this context reduced to a more intimate micro-scale. Nightingale's foundational text *Notes on Nursing: What it is and What it is Not* (Nightingale 1859) was intended to create useful clinical knowledge on health care settings and work. Specifically, it considered the conditions, variety, arrangements, behaviors and interactions in settings, and how these might be manipulated to create better care and health (Andrews 2003; Selanders 1998). In particular, *Notes* considered the importance of micro-environmental conditions such as ventilation, warmth and light and their microsocial conditions such as nurses' proximities to and interactions with patients.

As a key historical and extremely influential text, *Notes* warrants some dedicated attention and deconstruction here. Nightingale's introduction to *Notes* argued that human suffering may not always be associated with universal symptoms, but may often be brought about, or at least contributed to significantly, by a patient's immediate environmental conditions. She commented:

> In watching disease, both in private houses and in public hospitals, the thing which strikes the experienced observer most forcibly is this, that the symptoms or the sufferings generally considered to be inevitable and incident to the disease are very often not symptoms of the disease at all, but of something quite different—of the want of fresh air, or of light, or of warmth, or of quiet, or of cleanliness, or of punctuality and care in the administration of diet, of each or all or these (1859, p. 5).

Nightingale argued that the term 'nursing', should be broadened from its narrow definition, to be concerned with these important environmental factors. In discussing living conditions, Nightingale focused on houses as specific health care settings.

She listed five essential conditions for securing the health of houses: pure air, pure water, efficient drainage, cleanliness and light. She thought that if addressed, and enhanced individually, they would reduce illness and suffering. Referring to fresh air for example, she suggested:

> The very first canon of nursing, the first and the last thing upon which a nurses attention must be fixed, the first essential to the patient, without which all the rest you can do for him is as nothing... is: TO KEEP THE AIR HE BREATHES AS PURE AS THE EXTERNAL AIR, WITHOUT CHILLING HIM. (original emphasis) (1859, p. 8)

Nightingale then describing specific tactics for ventilation, explaining:

> With a proper supply of windows, and a proper supply of fuel in open fire places, fresh air is comparatively easy to secure when your patient or patients are in bed. Never be afraid of opening windows then. People don't catch cold in bed ... But a careless nurse ... will stop up every cranny and keep a hot house heat ... (1859, p. 9)

Laying an important basis for the 'rules' of personal care, the remainder of *Notes* is concerned with various features of what Nightingale termed the 'petty management of illness'—'how to manage that what you do in a place'. Again, a substantial proportion of her observations and recommendations were concerned with the patient's immediate physical environment and the management of it, while others were focused more on nurses' caring actions. Her first set of observations and recommendations concerned noise and bodily movement. For example, Nightingale advised, for the sake of patients, to avoid external and excessive noise, to whisper conversations in a patient's room and to be quiet when standing directly outside of it. She commented on everyday bodily movement around patients and on the downfalls of the period's latest female fashions and associated body rituals:

> Compelled by her dress, every woman now either shuffles or waddles—only a man can cross the floor of a sick room without shaking it! What is become of woman's light step?—the firm, light quick step we have been asking for?... The nurse who rustles is the horror of a patient, though perhaps he does not know why. The fidget of silk and of crinoline, the rattling of keys, the creaking of stays and of shoes, will do a patient more harm than all the medicines in the world will do him good (1859, 27)

Nightingale also warned against rapid body movements, such as hurrying and rushing. She regarded subtle body positioning within a room to be of paramount importance to the patient's welfare, instructing:

> Always sit down when a sick person is talking business to you ... by continuing to stand you make him raise his eyes to see you ... Always sit within the patient's view, so that when you speak to him, he has not painfully to turn his head round in order to look at you (1859, 28)

Continuing with the management of illness, Nightingale also considered the importance, and various features, of what she termed 'variety' in a patient's immediate physical environment. This was motivated by her concern over the monotony and boredom created by a patient's observation of the same dull regular surroundings over long durations, and its concurrent impact on their mental and physical health. She stated:

> To any but an old nurse, or an old patient, the degree would be quite inconceivable to which the nerves of the sick suffer from seeing the same walls, the same ceiling, the same surroundings, during long confinements (1859, p. 33)

As remedies to boring surroundings, Nightingale suggested a number of human interventions including the provision of beautiful objects (what she termed 'fancies') and subtle spatial changes to a room, such as moving the position of the patient's bed. These were combined with therapeutic nature-orientated interventions such as allowing the patient to look out of a window regularly and the giving of flowers.

In comparison to contemporary nursing research, little attention is given in Nightingale's work to a reciprocal nurse–patient relationship though she did discuss the benefits of oral communication, and of 'chattering hope and advices' (Andrews 2003). In sum, its complete lack of contemporary liberal sensitivities acknowledged, *Notes* and its comprehensive micro-scale analysis influences nursing research, teaching and practice to this day.

The Metaparadigm Concept of 'Nursing Environment' (Twentieth Century)

'Nursing environment' re-emerged in the mid-twentieth century at the same time as nursing research and training was becoming established in universities, specifically as one of the four metaparadigm concepts of nursing theory at the time: *nursing, person, health, environment* (see Thorne et al. 1998; Watson 1979). The fundamental question posed by nurse theorists was 'what is nursing environment?' In terms of answers, as Fitzpatrick and Whall (1983) describe, a common distinction was made in much literature between aspects of, and occurrences with and within, nurses' bodies (thought of as 'internal' events) and everything else beyond the body (thought of 'external' events). Specifically, three progressively increasing degrees of proximity and interaction between internal and external events representing three understandings of nursing environment. Indeed, one understanding held internal and external events apart, internal for example being the nurse and external being other human and physical contexts (Peplau 1952a, b; Orlando 1961). A second understanding brought internal and external events closer together, the exchange between the two—such as personal physiology/psychology *with* health system and societal entities—being highly important (Levine 1969; Patterson and Zderad 1976; Neuman 1980). A third understanding more fully integrated internal and external events, stressing their permeable boundaries, constant interplay, interdependence and exchanges of energy, matter and information (Roy 1976; Rogers 1980; King 1981; Parse 1981; Quinn 1992). Of this latter group, *Roy's Adaptation Model* of Nursing is a particularly influential example (see Roy 1976). Here the patient is conceptualized as a series of systems—biological, psychological and social (i.e. a 'biopsychosocial' person) who is constantly attempting to adapt and maintain the balance between themselves and the outside world to maximize their own coping and happiness; the nurse's role being to promote this adaptation and maintenance. According to Roy, three types of environmental stimuli are critical to both the nurse's and patient's ability to do this: (1) 'focal' (that immediately confront and individual—such as their needs or consequences of their symptoms); (2) 'contextual' (all other external influences—such as past experiences, values, culture); and (3) 'residual' (similar to contextual but are unclear or speculated).

In sum, the concept of nursing environment is a key part of a general intellectual project of its time to find an explanatory grand theory of nursing to help support, justify, secure and build a legitimate and distinct profession and academic discipline (Andrews 2016). Such 'big theory', however, has gradually given way in recent years—particularly in the context of changing and more favorable academic and institutional contexts for nurses—to more fragmented, specialized and critical theorization of particular nursing attributes, agency and responsibilities, this itself energized by the increasing number of nurse academics with training in the social sciences and social theory (Andrews 2016).

'Practice Environments' of Various Forms and Scales in Contemporary Research (Late Twentieth and Early Twenty-First Centuries)

In recent decades, although there has been a reducing emphasis on nursing environment as an overarching concept, other ideas and interests have replaced it, most of which fall under the broad multidisciplinary and multi-scaled notion of 'practice environments'. The proliferation of health professional research means that countless hundreds of published papers fall in some way under it, a number of key themes emerging:

1. A prominent topic is 'work environments' or 'workplace issues' including the support, empowerment and opportunities they offer (Almost and Spence-Laschinger 2002; Fallatah and Laschinger 2016; Haugh and Laschinger 1996; Regan et al. 2016; Tourangeau et al. 2009), and their social functioning and what makes them psychologically or morally healthy or unhealthy (Dendaas 2004, 2010; Hayes et al. 2015; Lavoie-Tremblay 2004; Lavoie-Tremblay et al. 2008; Leveck and Jones 1996; Park et al. 2015; Peter et al. 2004; Shirey 2006; Vanderheide et al. 2013; Vessey et al. 2009). Specifically bringing clients more broadly into considerations of institutional work environments is the topic of 'ward atmospheres' and how their interpersonal (Brunt and Rask 2007; Caldwell et al. 2006; Corey et al. 1986; Eklund and Hansson 2001; Friis 1986; Middelboe et al. 2001), and architectural and design features (Becker 2007; Parker et al. 2004; Rashid 2006; Marquardt and Schmieg 2009; Whittaker and Chee 2015; Williams 2001) impact on the functioning of care, as well as on therapeutic experiences and outcomes.

2. Practice environment has been mobilized in debates on the 'best places to care', institutions versus homes (Björnsdóttir et al. 2015; Parratt and Fahy 2004; Smith-Stoner 2002; West et al. 2000; Watty et al. 2003) being a popular topic given evolving policies and priorities over the years. A specific interest has also emerged in the idea of 'magnet' hospitals and health care settings (Buchan 1999; Kramer and Schmalenberg 1988; McCaughey et al. 2020; Scott et al. 1999; Upenieks 2003a, b). This an idea forwarded in order to determine what can be learned from the most high-performing attractive, popular and successful places. Particular magnet qualities are said to include transformational leadership and innovation, low staff turnover and high staff satisfaction, high staff involvement and structured empowerment, top patient outcomes and open communication; magnet status is even an official certification in certain counties and jurisdictions.
3. Practice environment has been mobilized in debates on the 'best ways to care'. Debates have continued on the nature and importance of physical 'co-presence' in caring relationships (Melnechenko 2003; Osterman and Schwartz-Barcott 1996; MacKinnon et al. 2005); this being particularly important given the emergence of caring in and through the internet in the last 20 years (Baldassar 2016; Cudney and Weinert 2000; Hern et al. 1997). Otherwise debates have emerged on leadership potential in practice environments for practice enhancement and development (McCormack and McCance 2011; McCormack et al. 2013), oftentimes practice environments regarded as 'contexts' with political, economic and social variability that help or hinder research implementation/knowledge translation (Kitson et al. 1998, 2008; Rycroft-Malone 2004; Rycroft-Malone et al. 2013; Yost et al. 2015).
4. Practice environment has been conceived in many ways 'beyond the hospital and bedside'. Urbanicity and rurality (Baernholdt and Mark 2009; Duggleby et al. 2016; Long and Weinert 2006; Pauly 2014), cyberspace (Andrew et al. 2012; Birchley et al. 2017; Cudney and Weinert 2000), local determinants of client health (Muhlenkamp et al. 1985; Williams et al. 2014), therapeutic communities and mileus (Jain et al. 2016; Thomas et al. 2002; Tuck and Keels 1992) all becoming popular focuses of research attention. In this vein, the 'natural environment' has emerged as a common area of contemporary interest (see Chinn 1996; Kleffel 1991; Schuster and Brown 1994), much of this work being focused on pollution and 'environmental health' as a growing practice concern (Neufer 1994; Grady et al. 1997; Larsson and Butterfield 2002; Severtson et al. 2002; Van Dongen 2002; Butterfield 2002; Dixon and Dixon 2002; Wright 2003; Sattler and Lipscomb 2003; McCurdy et al. 2004; Watterson et al. 2005; Sweeney and de Peyster 2005).
5. Moving up in scale, the 'region and nation' have become recognised as important practice contexts. Nurse scholars, for example, often talk about the development of their national or regional context with regard to a predominant culture of, or way of doing, nursing (Buljac-Samardzic et al. 2016; Gill 2004; Kotzer and Arellana 2008; Leininger 1994; Marks 1994), or overall developments, roles, capacities and resources in this context (Bressan et al. 2016; Bramadat and

Chalmers 1989; Scribante and Bhagwanjee 2007; DiCenso et al. 2010). There is also an appreciation of the historical development of professional practice, research telling alternative, often overlooked, histories rooted in practices and places that lie outside and challenge dominant western narratives. Indeed, this approach is often focused on nursing at the geographical margins of specific regions and nations (Arseneault 2008; Burnett 2008; Daigle 2008; Elliott et al. 2008; Elliott 2008; Nelson 2001; Majda et al. 2015; McKay 2008; Quiney 2008; Rutherdale 2008; Searle 1965; Stuart 2008; Toman 2008; Timmons et al. 2016).

6. Linking with, contributing to, the broad field of global health, at the very largest of scales, 'the global' has become a focus of critical attention both as a context for practice and perhaps even a practice environment in its own right. This is particularly the case in nursing research, which considers such subjects as the identity of a collective global community of nurses (Falk-Rafael 2006; Wilson et al. 2016); nurse migrations and the global health care economy (Aluttis et al. 2014; Buchan 2002; Kingma 2006; Marchal and Kegels 2003; Squires et al. 2016); global health diseases, issues and challenges for nurses and clinicians (Bateman et al. 2001; Carlton et al. 2007; Davis 1998; Rowson et al. 2012; von Strauss et al. 2017) and globalization processes more broadly as they impact upon the profession (Austin 2001; Baumann and Blythe 2008; Cartwright 2000; Wilson et al. 2016; Messias 2001; Wrede 2010) most notably including a focus on how it effects the world ecosystem and climate change (Andrews 2009; Kleffel 1996; Kirk 2002; Laustsen 2006; Richardson et al. 2016).

Expansive and Personal Horizons: From 'The World' to 'Lifeworlds' in Contemporary Practice (Late Twentieth and Early Twenty-First Centuries)

Finally, although not always an explicit focus of attention, various uses and understandings of the geographical meta-concept of 'world' are common in health professional research (these emerging in parallel to the aforementioned ideas on clinical environment). Two main variations exist.

1. The most general and frequent use of world is 'The World'. 'The World' meaning all earthly existence, the total surround within which humans are situated. Across research, the entirety of 'The World'—the 'whole world'—is often emphasized—such as 'nursing in today's world' (Ellis and Hartley 2004; Hartley 1988). In practice, however, mention of the whole world in research is almost always in the context of specific/narrower thematic content. Oftentimes, for example, particular phenomenon that prevail differentially 'around the world' are considered, such as different practice ethics (Ludwick and Silva 2000) or views on disease (Navon 1999). Scholars meanwhile have looked towards world more holistically in the context of promoting a particular expansive vision and

outlook, such as calling for 'worldly thinking' and approaches in practice (McIntyre 2003).

'The World' frequently arises practically in the titles of institutions and institutional initiatives, that health professionals and researchers inevitably have to engage with if they hope to understand important structural influences on health care operating at the very largest of scales. Examples include studying the priorities, operations, data and outcomes of the *World Bank, World Trade Organization, World Health Organization* and *One World Health* (Martin and Newcombe 2001; Messias 2001; WHO 2016; Wong et al. 2015). Ironically, 'The World' is often purposefully divided by such institutions along extensive lines, one side of the divide providing a contrast with the other. Such divides might be social and economic in nature, and picked up by research such as 'third world' vs 'first world' practice, 'developing world' vs 'developed world' practice or 'low income' vs 'high income' country practice (Aveling et al. 2015; Board 2001; Haloburdo and Thompson 1998; Thompson et al. 2000); there being debate over their positioning, resulting inclusions/exclusions, meaning attributions and resource implications.

Another deployment of 'The World' is in the context of world-level changes. The everyday term 'changing world' is frequently used to refer to ubiquitous political, economic and social developments, oftentimes this being a backdrop for a specific topic of interest, such as 'nursing in a changing world' (Scott and West 2001). Other times, however, what has changed with 'The World' is more specifically acknowledged in research, studies considering, for example, nursing in a 'technological world' (Marck 2000) or nursing in an 'aging world' (Swanson 1999). Research can also be speculative and visionary, about events and worlds that have not yet happened. If those futures are preferred research might radiate an affective sense of hope and possibility for how they might be and feel, such as imagining and planning practice for a 'post-COVID-19 world' (Tingle 2020). Although of course those futures might be unwanted, research then radiating an affective apprehension of how they will be and feel, such as imaging and planning for practice after climate change (Leyva et al. 2017). Indeed, in much literature there is at least some sense of new worlds emerging from events.

2. The other variation of world in health professional research the enduring idea of 'lifeworld'. Lifeworld has roots in human geography and phenomenological philosophy. It is based upon human experience and the meanings human surrounds accrue for them through their presence and involvement in them, particularly through specific sets of cultural or economic practices. Socially and spatially, then, a lifeworld is the total purview and existence of an individual or group as it is subjectively felt and known. It is, if you like, their particular lived slice of 'The World.' A number of uses of and variations on lifeworld are apparent in the health professional literature.

   Lifeworld has been promoted across as a humanistic concept to frame better caring practice and relationships (Galvin and Todres 2013; Hörberg et al. 2011; Todres et al. 2007), for example with regard to nursing both generally (Benner 2000; Dahlberg and Drew 1997; Lyckhage et al. 2018) and its specific clinical specialties (Biley and Galvin 2007; Ekebergh 2007). In this literature, the working domains of health professionals are considered lifeworlds as are the lived domains of clients, the meetings between the two being of interest. At these intersections, lifeworld helps understand the nature and immediacy of professional-client encounters—including realms of responsibility/agency and emotional entanglements—with the aim to maximise open-mindedness and understanding. In short, this work hopes to assure that under unequal power relationships professional lifeworlds do not in the moment unnecessarily ignore, disrespect or dominate over clients' lifeworlds. Notably, recently emerging has been the aligned idea of 'system worlds' (or system lifeworlds) lived by clients. They arise as they move through and between different institutional settings over time, their particular symbols and activities producing a long-term institutionalized lived experience of illness and care (Brown 2011; Froggatt et al. 2011).

Across research, a general concern also arises for 'the lived world of…' whereby entities are thought to create their own lifeworlds, either through providing a broad set of practices and experiences that an individual might dip into and out of, or through consuming and enveloping the individual in a particular intense personal experience. With regards to the former, we see attention in studies for example to lived worlds of specific forms of health care work (Giudici and Morselli 2019) and health research work (Brisbois and Plamondon 2018). With regards to the latter, we see attention in studies to the lived world of certain health conditions, which diminish and replace sufferers' former lifeworlds, providing a different reality, for example, the lived world of tuberculosis (Bender et al. 2010). Acknowledged, however, is how some treatments, care and assistive technologies have the potential to at least partially restore client's former lifeworlds (Pettersson et al. 2005, 2007).

A general acknowledgement exists that different worlds/lifeworlds arise relationally. Research for example differentiates client's internal family/cultural worlds and the 'outside world' (Rousseau et al. 2009), and the 'parallel worlds' of social/community life and institutionally-based biomedicine (Arnold et al. 2018). Studying this literature, there seems to be an implicit phenomenological understanding that the specificities of lifeworlds, and the meaning-making experiences of life lived in them, often leads to the development of individual and group 'worldviews'. These as sets of beliefs and understandings on those lifeworlds and worlds beyond. Worldviews might act as an extension/projection of an individual or group's lifeworlds, potentially clashing or complementing or influencing other lifeworlds and worldviews subsequently encountered by them. This often occurs for example in the case of holistic/traditional health worldviews vs biomedical worldviews (Arnold and Bruce 2005; Boon 1998; Hibbert et al. 2003; Martsolf and Mickley 1998).

Summary

This chapter has run through the long and rich history of geographical thinking in health care practice, tracing origins in concerns for nature and the natural, through to contemporary awareness of practice contexts. Many of the ideas in this chapter will be touched upon later in the book, and we are now well-placed to consider the most recent trends in geographical thinking in the remaining chapters.

References

Almost, J., & Spence-Laschinger, H. K. (2002). Workplace empowerment, collaborative work relationships, and job strain in nurse practitioners. *Journal of the American Academy of Nurse Practitioners, 14*(9), 408–420.

Aluttis, C., Bishaw, T., & Frank, M. W. (2014). The workforce for health in a globalized context–global shortages and international migration. *Global Health Action, 7*(1), 23611.

Anderson, W. P. (1998). Where is the postcolonial history of medicine? *Bulletin of the History of Medicine, 72*(3), 522–530.

Andrew, S., Cleary, M., & Jackson, D. (2012). Facing a new frontier: Safety in cyberspace and challenges for nursing. *Journal of Psychosocial Nursing and Mental Health Services, 50*(8), 4–5.

Andrews, G. J. (2003). Nightingale's geography. *Nursing Inquiry, 10*(4), 270–274.

Andrews, G. J. (2009). Global climate change: A framework for nursing action. *Aporia, 1*(1), 22–29.

Andrews, G. J. (2016). Geographical thinking in nursing inquiry, part one: Locations, contents, meanings. *Nursing Philosophy, 17*(4), 262–281.

Arnold, O. F., & Bruce, A. (2005). Nursing practice with aboriginal communities: Expanding worldviews. *Nursing Science Quarterly, 18*(3), 259–263.

Arnott, R. (1996). Healing and medicine in the Aegean Bronze Age. *Journal of the Royal Society of Medicine, 89*(5), 265–270.

Arnold, R., Van Teijlingen, E., Ryan, K., & Holloway, I. (2018). Parallel worlds: An ethnography of care in an Afghan maternity hospital. *Social Science & Medicine, 216*, 33–40.

Arseneault, A. (2008). University nursing education for francophones in New Brunswick in the 1960s: The role of nuns, priests, politicians, and nurses. In J. Elliott, M. E. Stuart, & C. Toman (Eds.), *Place and practice in Canadian nursing history*. Vancouver: UBC Press.

Austin, W. (2001). Nursing ethics in an era of globalization. *Advances in Nursing Science, 24*(2), 1–18.

Aveling, E. L., Kayonga, Y., Nega, A., & Dixon-Woods, M. (2015). Why is patient safety so hard in low-income countries? A qualitative study of healthcare workers' views in two African hospitals. *Globalization and Health, 11*(1), 1–8.

Baernholdt, M., & Mark, B. A. (2009). The nurse work environment, job satisfaction and turnover rates in rural and urban nursing units. *Journal of Nursing Management, 17*(8), 994–1001.

Baldassar, L. (2016). Mobilities and communication technologies: Transforming care in family life. In M. Kilkey & E. Palenga-Möllenbeck (Eds.), *Family life in an age of migration and mobility* (pp. 19–42). London: Palgrave Macmillan.

Barrett, F. A. (1991). 'Scurvy' Lind's medical geography. *Social Science & Medicine, 33*(4), 347–353.

Barrett, F. A. (1996). Daniel Drake's medical geography. *Social Science & Medicine, 42*(6), 791–800.

Barrett, F. A. (1998). Alfred Haviland's nineteenth-century map analysis of the geographical distribution of diseases in England and Wales. *Social Science & Medicine, 46*(6), 767–781.

Barrett, F. A. (2000a). *Disease & geography: The history of an idea*. Toronto: Atkinson College Department of Geography, York University.

Barrett, F. A. (2000b). August Hirsch: As critic of, and contributor to, geographical medicine and medical geography. *Medical History Supplement, 20*, 98.

Barrett, F. A. (2000c). Finke's 1792 map of human diseases: the first world disease map?. *Social Science & Medicine, 50*(7), 915–921.

Barrett, F. A. (2002). The role of French-language contributors to the development of medical geography (1782–1933). *Social Science & Medicine, 55*(1), 155–165.

Bateman, C., Baker, T., Hoornenborg, E., & Ericsson, U. (2001). Bringing global issues to medical teaching. *The Lancet, 358*(9292), 1539–1542.

Baumann, A., & Blythe, J. (2008). Globalization of higher education in nursing. *The Online Journal of Issues in Nursing, 13*(2), 4.

Becker, F. (2007). Nursing unit design and communication patterns: What is "real" work? *Health Environments Research & Design Journal, 1*(1), 58–62.

Bender, A., Andrews, G., & Peter, E. (2010). Displacement and tuberculosis: Recognition in nursing care. *Health & Place, 16*(6), 1069–1076.

Benner, P. (2000). The roles of embodiment, emotion and lifeworld for rationality and agency in nursing practice. *Nursing Philosophy, 1*(1), 5–19.

Biley, F. C., & Galvin, K. T. (2007). Lifeworld, the arts and mental health nursing. *Journal of Psychiatric and Mental Health Nursing, 14*(8), 800–807.

Birchley, G., Huxtable, R., Murtagh, M., Ter Meulen, R., Flach, P., & Gooberman-Hill, R. (2017). Smart homes, private homes? An empirical study of technology researchers' perceptions of ethical issues in developing smart-home health technologies. *BMC Medical Ethics, 18*(1), 1–13.

Björnsdóttir, K., Ceci, C., & Purkis, M. E. (2015). The 'right' place to care for older people: Home or institution? *Nursing Inquiry, 22*(1), 64–73.

Boon, H. (1998). Canadian naturopathic practitioners: Holistic and scientific world views. *Social Science & Medicine, 46*(9), 1213–1225.

Board, S. (2001). Risks to health care workers in developing countries. *N Engl j Med, 345*(7), 3–14.

Bramadat, I. J., & Chalmers, K. I. (1989). Nursing education in Canada: Historical 'progress'—contemporary issues. *Journal of Advanced Nursing, 14*(9), 719–726.

Bressan, V., Tolotti, A., Barisone, M., Bagnasco, A., Sasso, L., Aleo, G., et al. (2016). Perceived barriers to the professional development of modern nursing in Italy—A discussion paper. *Nurse Education in Practice, 17*, 52–57.

Brisbois, B., & Plamondon, K. (2018). The possible worlds of global health research: An ethics-focused discourse analysis. *Social Science & Medicine, 196*, 142–149.

Brown, P. (2011). The concept of lifeworld as a tool in analysing health-care work: Exploring professionals' resistance to governance through subjectivity, norms and experiential knowledge. *Social Theory & Health, 9*(2), 147–165.

Brown, T., & Moon, G. (2004). From Siam to New York: Jacques May and the 'foundation' of medical geography. *Journal of Historical Geography, 30*(4), 747–763.

Brunt, D., & Rask, M. (2007). Ward atmosphere—The Scarlet Pimpernel of psychiatric settings? *Issues in Mental Health Nursing, 28*(6), 639–655.

Buchan, J. (1999). Still attractive after all these years? Magnet hospitals in a changing health care environment. *Journal of Advanced Nursing, 30*(1), 100–108.

Buchan, J. (2002). Global nursing shortages: Are often a symptom of wider health system or societal ailments. *BMJ: British Medical Journal, 324*(7340), 751.

Buljac-Samardzic, M., van Wijngaarden, J. D., & Dekker-van Doorn, C. M. (2016). Safety culture in long-term care: A cross-sectional analysis of the Safety Attitudes Questionnaire in nursing and residential homes in the Netherlands. *BMJ Quality & Safety, 25*(6), 424–431.

Burnett, K. (2008). The healing work of Aboriginal women in indigenous and newcomer communities. In J. Elliott, M. E. Stuart, & C. Toman (Eds.), *Place and practice in Canadian nursing history*. Vancouver: UBC Press.
Butterfield, P. G. (2002). Upstream reflections on environmental health: An abbreviated history and framework for action. *Advances in Nursing Science, 25*(1), 32–49.
Caldwell, B. A., Gill, K. J., Fitzgerald, E., Sclafani, M., & Grandison, P. (2006). The association of ward atmosphere with burnout and attitudes of treatment team members in a state psychiatric hospital. *American Journal of Psychiatric Rehabilitation, 9*(2), 111–129.
Carlton, K. H., Ryan, M., Ali, N. S., & Kelsey, B. (2007). Integration of global health concepts in nursing curricula: A national study. *Nursing Education Perspectives, 28*(3), 124–129.
Cartwright, L. (2000). Reach out and heal someone: Telemedicine and the globalization of health care. *Health, 4*(3), 347–377.
Clemow, F. G. (1903). The geography of disease. Cambridge University Press.
Chinn, P. L. (1996). Environment, health and nursing. *Advances in Nursing Science, 18*(4), 48–59.
Corey, L. J., Wallace, M. A., Harris, S. H., & Casey, B. (1986). Psychiatric ward atmosphere. *Journal of Psychosocial Nursing and Mental Health Services, 24*(10), 10–16.
Crozier, A. (2007). *Practising colonial medicine: The colonial medical service in British East Africa*. London: IB Tauris.
Cudney, S., & Weinert, C. (2000). Computer-based support groups: Nursing in cyberspace. *Computers in Nursing, 18*, 35–46.
Dahlberg, K., & Drew, N. (1997). A lifeworld paradigm for nursing research. *Journal of Holistic Nursing, 15*(3), 303–317.
Daigle, J. (2008). The call of the north: Settlement nurses in the remote areas of Québec, 1932-72. In J. Elliott, M. E. Stuart, & C. Toman (Eds.), *Place and practice in Canadian nursing history*. Vancouver: UBC Press.
Davis, D. (1998). Continuing medical education. Global health, global learning. *British Medical Journal, 316*, 385–386.
Dendaas, N. (2004). The scholarship related to nursing work environments: Where do we go from here? *Advances in Nursing Science, 27*(1), 12–20.
Dendaas, N. (2010). Environmental congruence and work-related stress in acute care hospital medical/surgical units: A descriptive, correlational study. *HERD, 5*(1), 23–42.
DiCenso, A., Martin-Misener, R., Bryant-Lukosius, D., Bourgeault, I., Kilpatrick, K., Donald, F., et al. (2010). Advanced practice nursing in Canada: Overview of a decision support synthesis. *Nursing Leadership, 23*, 15–34.
Dixon, J. K., & Dixon, J. P. (2002). An integrative model for environmental health research. *Advances in Nursing Science, 24*(3), 43–57.
Drake D (1850). A systematic treatise, historical, etiological and practices on the principles of disease. Vol 1 American classics in history and social sciences. New York.
Drake D (1854). A systematic treatise, historical, etiological and practices on the principles of disease. Vol 2 American classics in history and social sciences. New York.
Duggleby, W., Robinson, C. A., Kaasalainen, S., Pesut, B., Nekolaichuk, C., MacLeod, R., et al. (2016). Developing navigation competencies to care for older rural adults with advanced illness. *Canadian Journal on Aging, 35*(2), 206–214.
Ekebergh, M. (2007). Lifeworld-based reflection and learning: A contribution to the reflective practice in nursing and nursing education. *Reflective Practice, 8*(3), 331–343.
Eklund, M., & Hansson, L. (2001). Ward atmosphere, client satisfaction, and client motivation in a psychiatric work rehabilitation unit. *Community Mental Health Journal, 37*(2), 169–177.
Elliott, J. (2008). (Re)constructing the identity of a red cross outpost nurse: The letters of Louise de Kiriline, 1927-36. In J. Elliott, M. E. Stuart, & C. Toman (Eds.), *Place and practice in Canadian nursing history*. Vancouver: UBC Press.
Elliott, J., Stuart, M. E., & Toman, C. (2008). *Place and practice in Canadian nursing history*. Vancouver: UBC Press.

Ellis, J. R., & Hartley, C. L. (2004). *Nursing in today's world: Trends, issues & management.* Philadelphia: Lippincott Williams & Wilkins.

Ernst, W. (2007). Beyond east and west. From the history of colonial medicine to a social history of medicine (s) in south Asia. *Social History of Medicine, 20*(3), 505–524.

Falk-Rafael, A. (2006). Globalization and global health: Toward nursing praxis in the global community. *Advances in Nursing Science, 29*(1), 2–14.

Fallatah, F., & Laschinger, H. K. (2016). The influence of authentic leadership and supportive professional practice environments on new graduate nurses' job satisfaction. *Journal of Research in Nursing, 21*(2), 125–136.

Fitzpatrick, J., & Whall, A. (1983). *Conceptual models of nursing: Analysis and application.* Upper Saddle River: Prentice-Hall.

Friis, S. (1986). Characteristics of a good ward atmosphere. *Acta Psychiatrica Scandinavica, 74*(5), 469–473.

Froggatt, K., Hockley, J., Parker, D., & Brazil, K. (2011). A system lifeworld perspective on dying in long term care settings for older people: Contested states in contested places. *Health & Place, 17*(1), 263–268.

Galvin, K., & Todres, L. (2013). *Caring and well-being: A lifeworld approach.* Milton Park: Routledge.

Gill, P. (2004). Difficulties in developing a nursing research culture in the UK. *British Journal of Nursing, 13*(14), 876–879.

Giudici, F., & Morselli, D. (2019). 20 Years in the world of work: A study of (nonstandard) occupational trajectories and health. *Social Science & Medicine, 224*, 138–148.

Grady, P., Harden, J. T., Mortiz, P., & Amende, L. M. (1997). Incorporating environmental sciences and nursing research: An NINR initiative. *Nursing Outlook, 45*(2), 73–75.

Haloburdo, E. P., & Thompson, M. A. (1998). A comparison of international learning experiences for baccalaureate nursing students: Developed and developing countries. *Journal of Nursing Education, 37*(1), 13–21.

Hartley, C. L. (1988). *Nursing in today's world: Challenges, issues, and trends.* Philadelphia: Lippincott Williams & Wilkins.

Harvard (2017). Airs, waters and places: a climate change series. Harvard Public Health Magazine.

Haugh, F. B., & Laschinger, H. S. (1996). Power and opportunity in public health nursing work environments. *Public Health Nursing, 13*(1), 42–49.

Haviland, A. (1855). Climate, Weather, and Disease: Being a Sketch of the Opinions of the Most Celebrated Antient and Modern Writers. J. Churchill.

Haviland, A. (1892). The geographical distribution of disease in Great Britain. Swan Sonnenschein.

Hayes, B., Douglas, C., & Bonner, A. (2015). Work environment, job satisfaction, stress and burnout among haemodialysis nurses. *Journal of Nursing Management, 23*(5), 588–598.

Hern, M., Weitkamp, T., Haag, D., Trigg, J., & Guard, J. (1997). Nursing the community in cyberspace. *Computers in Nursing, 15*, 316–323.

Hibbert, D., Hanratty, B., May, C., Mair, F., Litva, A., & Capewell, S. (2003). Negotiating palliative care expertise in the medical world. *Social Science & Medicine, 57*(2), 277–288.

Hörberg, U., Ozolins, L. L., & Ekebergh, M. (2011). Intertwining caring science, caring practice and caring education from a lifeworld perspective—Two contextual examples. *International Journal of Qualitative Studies on Health and Well-Being, 6*(4), 10363.

Jain, S., McLean, C., Adler, E. P., & Rosen, C. S. (2016). Peer support and outcome for veterans with posttraumatic stress disorder (PTSD) in a residential rehabilitation program. *Community Mental Health Journal, 52*(8), 1089–1092.

Julien Viray (1817). Geographie Medicale. Dictionnaire des Sciences Medicales, 151–217..

Kearns, R., & Moon, G. (2002). From medical to health geography: Novelty, place and theory after a decade of change. *Progress in Human Geography, 26*(5), 605–625.

Kearns, R. A. (1993). Place and health: Towards a reformed medical geography. *The Professional Geographer, 45*(2), 139–147.

King, I. M. (1981). *A theory for nursing.* New York: Wiley.

Kingma, M. (2006). *Nurses on the move: Migration and the global health care economy*. Ithica: Cornell University Press.
Kirk, M. (2002). The impact of globalization and environmental change on health: Challenges for nurse education. *Nurse Education Today, 22*(1), 60–71.
Kitson, A., Harvey, G., & McCormack, B. (1998). Enabling the implementation of evidence based practice: A conceptual framework. *Quality in Health Care, 7*(3), 149–158.
Kitson, A., Rycroft-Malone, J., Harvey, G., McCormack, B., Seers, K., & Titchen, A. (2008). Evaluating the successful implementation of evidence into practice using the PARiHS framework: Theoretical and practical challenges. *Implementation Science, 3*(1), 1.
Kleffel, D. (1991). An ecofeminist analysis of nursing knowledge. *Nursing Forum, 26*(4), 5–18.
Kleffel, D. (1996). Environmental paradigms: Moving toward an ecocentric perspective. *Advances in Nursing Science, 18*(4), 1–10.
Kotzer, A. M., & Arellana, K. (2008). Defining an evidence-based work environment for nursing in the USA. *Journal of Clinical Nursing, 17*(12), 1652–1659.
Kramer, M., & Schmalenberg, C. (1988). Magnet hospitals: Part I. Institutions of excellence. *The Journal of Nursing Administration, 18*(1), 13–24.
Larsson, L. S., & Butterfield, P. (2002). Mapping the future of environmental health and nursing: Strategies for integrating national competencies into nursing practice. *Public Health Nursing, 19*(4), 301–308.
Laustsen, G. (2006). Environment, ecosystems, and ecological behavior: A dialogue toward developing nursing ecological theory. *Advances in Nursing Science, 29*(1), 43–54.
Laurent, E. (1905). Géographie médicale. Maloine.
Lavoie-Tremblay, M. (2004). Creating a healthy workplace: A participatory organizational intervention. *Journal of Nursing Administration, 34*(10), 469–474.
Lavoie-Tremblay, M., Wright, D., Desforges, N., Gélinas, C., Marchionni, C., & Drevniok, U. (2008). Creating a healthy workplace for new-generation nurses. *Journal of Nursing Scholarship, 40*(3), 290–297.
Leininger, M. (1994). The tribes of nursing in the USA culture of nursing. *Journal of Transcultural Nursing, 6*(1), 18–22.
Leveck, M. L., & Jones, C. B. (1996). The nursing practice environment, staff retention, and quality of care. *Research in Nursing & Health, 19*(4), 331–343.
Levine, M. E. (1969). *Introduction to clinical nursing*. Philadelphia: F.A. Davis.
Leyva, E. W. A., Beaman, A., & Davidson, P. M. (2017). Health impact of climate change in older people: An integrative review and implications for nursing. *Journal of Nursing Scholarship, 49*(6), 670–678.
Long, K. A., & Weinert, C. (2006). Rural nursing: Developing the theory base. In C. Winters (Ed.), *Rural nursing: Concepts, theory, and practice* (pp. 3–16). New York: Springer.
Ludwick, R., & Silva, M. C. (2000). Ethics: Nursing around the world: Cultural values and ethical conflicts. *Online Journal of Issues in Nursing, 5*(3), 1–10.
Lyckhage, E. D., Brink, E., & Lindahl, B. (2018). A theoretical framework for emancipatory nursing with a focus on environment and persons' own and shared lifeworld. *Advances in Nursing Science, 41*(4), 340–350.
MacKinnon, K., McIntyre, M., & Quance, M. (2005). The meaning of nurse's presence during childbirth. *Journal of Obstetric, Gynecologic, and Neonatal Nursing, 34*, 28–36.
Majda, A., Ziarko, E., & Zalewska-Puchała, J. (2015). A consistent course of events or a series of coincidences: Nursing in Poland from the 19th to the 21st century. *Nursing Inquiry, 22*(4), 359–370.
Marchal, B., & Kegels, G. (2003). Health workforce imbalances in times of globalization: Brain drain or professional mobility? *The International Journal of Health Planning and Management, 18*(S1), S89–S101.
Marck, P. (2000). Nursing in a technological world: Searching for healing communities. *Advances in Nursing Science, 23*(2), 62–81.

Marks, S. (1994). *Divided sisterhood: Race, class and gender in the South African nursing profession*. New York: Macmillan.
Marks, S. (1997). What is colonial about colonial medicine? And what has happened to imperialism and health? *Social History of Medicine, 10*(2), 205–219.
Marquardt, G., & Schmieg, P. (2009). Dementia-friendly architecture: Environments that facilitate wayfinding in nursing homes. *American Journal of Alzheimer's Disease and other Dementias, 24*(4), 333–340.
Martin, C., & Newcombe, E. A. (2001). Emergency care during the 1999 World Trade Organization meeting in Seattle. *Journal of Emergency Nursing, 27*(5), 478–480.
Martsolf, D. S., & Mickley, J. R. (1998). The concept of spirituality in nursing theories: Differing world-views and extent of focus. *Journal of Advanced Nursing, 27*(2), 294–303.
May, J. M. (1950). Medical geography: Its methods and objectives. *Geographical Review, 40*(1), 9–41.
May, J. M. (1959). *The ecology of human disease*. Springfield: MD Publications.
McCaughey, D., McGhan, G. E., Rathert, C., Williams, J. H., & Hearld, K. R. (2020). Magnetic work environments: Patient experience outcomes in Magnet versus non-Magnet hospitals. *Health Care Management Review, 45*(1), 21–31.
McCormack, B., Manley, K., & Titchen, A. (2013). *Practice development in nursing and healthcare*. Hoboken: John Wiley & Sons.
McCormack, B., & McCance, T. (2011). *Person-centred nursing: Theory and practice*. Chichester: John Wiley & Sons.
McCurdy, L. E., Roberts, J., Rogers, B., Love, R., Etzel, R., Paulson, J., et al. (2004). Incorporating environmental health into pediatric medical and nursing education. *Environmental Health Perspectives, 112*, 1755–1760.
McIntyre, M. (2003). Cultivating a worldly repose: The contribution of Sally Gadow's work to interpretive inquiry. *Nursing Philosophy, 4*(2), 111–120.
McKay, M. (2008). Region, faith, and health: The development of Winnipeg's Visiting Nursing Agencies, 1897-1926. In J. Elliott, M. E. Stuart, & C. Toman (Eds.), *Place and practice in Canadian nursing history*. Vancouver: UBC Press.
McLeod, K. S. (2000). Our sense of Snow: The myth of John Snow in medical geography. *Social Science & Medicine, 50*(7), 923–935.
Meade, M. S., & Earickson, R. (2000). *Medical geography*. New York: *Guildford Press*.
Melnechenko, K. (2003). To make a difference: Nursing presence. *Nursing Forum, 38*, 18–25.
Messias, D. K. H. (2001). Globalization, nursing, and health for all. *Journal of Nursing Scholarship, 33*(1), 9.
Middelboe, T., Schjødt, T., Byrsting, K., & Gjerris, A. (2001). Ward atmosphere in acute psychiatric in-patient care: Patients' perceptions, ideals and satisfaction. *Acta Psychiatrica Scandinavica, 103*(3), 212–219.
Mitman, G., & Numbers, R. L. (2003). From miasma to asthma: The changing fortunes of medical geography in America. *History and Philosophy of the Life Sciences, 25*(3), 391–412.
Muhlenkamp, A. F., Brown, N. J., & Sands, D. (1985). Determinants of health promotion activities in nursing clinic clients. *Nursing Research, 34*(6), 327–332.
Navon, L. (1999). Cultural views of cancer around the world. *Cancer Nursing, 22*(1), 39–45.
Nelson, S. (2001). *Say little, do much: Nursing, nuns, and hospitals in the nineteenth century*. Philadelphia: University of Pennsylvania Press.
Neufer, L. (1994). The role of the community health nurse in environmental health. *Public Health Nursing, 11*(3), 155–162.
Neuman, B. (1980). The Betty Neuman health-care systems model: A total person approach to patient problems. In J. P. Riehl & C. Roy (Eds.), *Conceptual models for nursing practice* (2nd ed.). New York: Appleton-Century-Crofts.
Nightingale, F. (1859). *Notes on nursing: What it is and what it is not*. London: Lippincott.
Noel Halle (1787). L'Air, ailments. Encyclopedie Medicine, 474–577.

Noel Halle (1792). Exposition du plan du'un traite complet d' hygiene. Le Medecine Sciences, 225–235.

Numbers, R. L. (2000). Medical science before scientific medicine: Reflections on the history of medical geography. *Medical History Supplement, 20*, 217–218.

Orlando, I. J. (1961). *The dynamic nurse–patient relationship*. New York: G.P. Putman's Sons.

Osterman, P., & Schwartz-Barcott, D. (1996). Presence: Four ways of being there. *Nursing Forum, 31*, 23–30.

Park, M., Cho, S. H., & Hong, H. J. (2015). Prevalence and perpetrators of workplace violence by nursing unit and the relationship between violence and the perceived work environment. *Journal of Nursing Scholarship, 47*(1), 87–95.

Parker, C., Barnes, S., McKee, K., Morgan, K., Torrington, J., & Tregenza, P. (2004). Quality of life and building design in residential and nursing homes for older people. *Ageing and Society, 24*(6), 941–962.

Parratt, J., & Fahy, K. (2004). Creating a 'safe' place for birth: An empirically grounded theory. *New Zealand College of Midwives Journal, 30*, 11–14.

Parse, R. R. (1981). *Man–living–health: A theory of nursing*. New York: Wiley.

Patterson, J. C., & Zderad, L. T. (1976). *Humanistic nursing*. New York: Wiley.

Paul, B. K. (1985). Approaches to medical geography: An historical perspective. *Social Science & Medicine, 20*(4), 399–404.

Pauly, B. (2014). Close to the street: Nursing practice with people marginalized by homelessness and substance use. In S. W. Hwang, M. Guirguis-Younger, & R. McNeil (Eds.), *Homelessness & health in Canada*. Ottawa: University of Ottawa Press.

Peplau, H. E. (1952a). *Interpersonal relations in nursing*. New York: Springer.

Peplau, H. E. (1952b). *Interpersonal relationships in nursing*. New York: G.P. Putman's Sons.

Peter, E., Macfarlane, A. V., & O'Brien-Pallas, L. L. (2004). Analysis of the moral habitability of the nursing work environment. *Journal of Advanced Nursing, 47*(4), 356–364.

Pettersson, I., Appelros, P., & Ahlström, G. (2007). Lifeworld perspectives utilizing assistive devices: Individuals, lived experience following a stroke. *Canadian Journal of Occupational Therapy, 74*(1), 15–26.

Pettersson, I., Berndtsson, I., Appelros, P., & Ahlström, G. (2005). Lifeworld perspectives on assistive devices: Lived experiences of spouses of persons with stroke. *Scandinavian Journal of Occupational Therapy, 12*(4), 159–169.

Phillips, D. R. (1985). Directions for medical geography in the 1980s: Some observations from the United Kingdom. *Social Science & Medicine, 20*(4), 404–407.

Quiney, L. (2008). "Suitable young women": Red Cross nursing pioneers and the crusade for healthy living in Manitoba, 1920-30. In J. Elliott, M. E. Stuart, & C. Toman (Eds.), *Place and practice in Canadian nursing history*. Vancouver: UBC Press.

Quinn, J. F. (1992). Holding sacred space: The nurse as healing environment. *Holistic Nursing Practice, 6*(4), 26–36.

Rashid, M. (2006). A decade of adult intensive care unit design: A study of the physical design features of the best-practice examples. *Critical Care Nursing Quarterly, 29*(4), 282–311.

Regan, S., Laschinger, H. K., & Wong, C. A. (2016). The influence of empowerment, authentic leadership, and professional practice environments on nurses' perceived interprofessional collaboration. *Journal of Nursing Management, 24*(1), E54–E61.

Richardson, J., Grose, J., Nelmes, P., Parra, G., & Linares, M. (2016). Tweet if you want to be sustainable: A thematic analysis of a Twitter chat to discuss sustainability in nurse education. *Journal of Advanced Nursing, 72*(5), 1086–1096.

Rogers, M. E. (1980). Nursing: A science of unitary man. In J. P. Riehl & C. Roy (Eds.), *Conceptual models for nursing practice* (2nd ed.). New York: Appleton-Century Crofts.

Rousseau, C., Hassan, G., Measham, T., Moreau, N., Lashley, M., Castro, T., et al. (2009). From the family universe to the outside world: Family relations, school attitude, and perception of racism in Caribbean and Filipino adolescents. *Health & Place, 15*(3), 751–760.

Rowson, M., Smith, A., Hughes, R., Johnson, O., Maini, A., Martin, S., et al. (2012). The evolution of global health teaching in undergraduate medical curricula. *Globalization and Health, 8*(1), 35.

Roy, C. (1976). *Introduction to nursing: An adaptation model*. Upper Saddle River: Prentice-Hall.

Rutherdale, M. (2008). Cleansers, cautious caregivers, and optimistic adventurers: A proposed typology of Arctic Canadian Nurses, 1945-70. In J. Elliott, M. E. Stuart, & C. Toman (Eds.), *Place and practice in Canadian nursing history*. Vancouver: UBC Press.

Rycroft-Malone, J. (2004). The PARIHS Framework—A framework for guiding the implementation of evidence-based practice. *Journal of Nursing Care Quality, 19*(4), 297–304.

Rycroft-Malone, J., Seers, K., Chandler, J., Hawkes, C. A., Crichton, N., & Allen C.& Strunin, L. (2013). The role of evidence, context, and facilitation in an implementation trial: Implications for the development of the PARIHS framework. *Implementation Science, 8*(1), 28.

Sattler, B., & Lipscomb, J. (2003). *Environmental health and nursing practice*. New York: Springer.

Schuster, E. A., & Brown, C. L. (1994). *Exploring our environmental connections*. Sudbury: Jones & Bartlett Publishers.

Scott, C., & West, E. (2001). Nursing in the public sphere: Health policy research in a changing world. *Journal of Advanced Nursing, 33*(3), 387–395.

Scott, J. G., Sochalski, J., & Aiken, L. (1999). Review of magnet hospital research: Findings and implications for professional nursing practice. *Journal of Nursing Administration, 29*(1), 9–19.

Scribante, J., & Bhagwanjee, S. (2007). National audit of critical care resources in South Africa-nursing profile. *South African Medical Journal, 97*(12), 1315–1318.

Searle, C. (1965). *The history of the development of nursing in South Africa, 1652-1960: A socio-historical survey*. Cape Town: Struik.

Selanders, L. C. (1998). The power of environmental adaptation: Florence Nightingale's original theory for nursing practice. *Journal of Holistic Nursing, 16*(2), 247–263.

Severtson, D. J., Baumann, L. C., & Will, J. A. (2002). A participatory assessment of environmental health concerns in an Ojibwa community. *Public Health Nursing, 19*(1), 47–58.

Shirey, M. R. (2006). Authentic leaders creating healthy work environments for nursing practice. *American Journal of Critical Care, 15*(3), 256–267.

Smith-Stoner, M. (2002). There's still no place like home: Results of the home healthcare nurse 2002 survey of nurses working in home care and hospice for 20+ years. *Home Healthcare Nurse, 20*, 657–662.

Spikins, P. (2015). *How compassion made us human*. Barnsley: Pen and Sword.

Spikins, P. (2017). Prehistoric origins: The compassion of far distant strangers. In P. Gilbert (Ed.), *Compassion: Concepts, research and applications*. Milton Park: Taylor & Francis.

Spikins, P., Needham, A., Tilley, L., & Hitchens, G. E. (2018a). Calculated or caring? Neanderthal healthcare in social context. *World Archaeology*. https://doi.org/10.1080/00438243.2018.1433060.

Spikins, P., Needham, A., Wright, B., Dytham, C., Gatta, M., & Hitchens, G. (2018b). Living to fight another day: The ecological and evolutionary significance of Neanderthal healthcare. *Quaternary Science Reviews*. https://doi.org/10.1016/j.quascirev.2018.08.011.

Spikins, P., Rutherford, H., & Needham, A. (2010). From homininity to humanity: Compassion from the earliest Archaics to modern humans. *Time & Mind: The Journal of Archaeology Consciousness and Culture, 3*(3), 303–325. https://doi.org/10.2752/175169610X12754030955977.

Squires, A., Ojemeni, M. T., & Jones, S. (2016). Exploring longitudinal shifts in international nurse migration to the United States between 2003 and 2013 through a random effects panel data analysis. *Human Resources for Health, 14*(1), 11–21.

Stuart, M. (2008). Social sisters: A feminist analysis of the discourses of Canadian Military Nurse Helen Fowlds, 1915-18. In J. Elliott, M. E. Stuart, & C. Toman (Eds.), *Place and practice in Canadian nursing history*. Vancouver: UBC Press.

Swanson, E. (1999). The world it is a-aging. *Journal of Gerontological Nursing, 25*(12), 35–37.

Sweeney, N. M., & de Peyster, I. (2005). Integrating environmental health into an undergraduate community health nursing course. *Public Health Nursing, 22*(5), 439–444.

Thomas, S. P., Shattell, M., & Martin, T. (2002). What's therapeutic about the therapeutic milieu? *Archives of Psychiatric Nursing, 16*(3), 99–107.

Thompson, K., Boore, J., & Deeny, P. (2000). A comparison of an international experience for nursing students in developed and developing countries. *International Journal of Nursing Studies, 37*(6), 481–492.

Thorne, S., Canam, C., Dahinten, S., Hall, W., Henderson, A., & Kirkham, S. (1998). Nursing's metaparadigm concepts: Disimpacting the debates. *Journal of Advanced Nursing, 27*(6), 1257–1268.

Timmons, S., Evans, C., & Nair, S. (2016). The development of the nursing profession in a globalised context: A qualitative case study in Kerala, India. *Social Science & Medicine, 166*, 41–48.

Tingle, J. (2020). Patient safety and litigation in the NHS post-COVID-19. *British Journal of Nursing, 29*(7), 444–445.

Todres, L., Galvin, K., & Dahlberg, K. (2007). Lifeworld-led healthcare: Revisiting a humanising philosophy that integrates emerging trends. *Medicine, Health Care and Philosophy, 10*(1), 53.

Toman, C. (2008). "A loyal body of empire citizens": Military nurses and identity at Lemnos and Salonika, 1915-17. In J. Elliott, M. E. Stuart, & C. Toman (Eds.), *Place and practice in Canadian nursing history*. Vancouver: UBC Press.

Tourangeau, A. E., Widger, K., Cranley, L. A., Bookey-Bassett, S., & Pachis, J. (2009). Work environments and staff responses to work environments in institutional long-term care. *Health Care Management Review, 34*(2), 171–181.

Tuck, I., & Keels, M. C. (1992). Milieu therapy: A review of development of this concept and its implicats for psychiatric nursing. *Issues in Mental Health Nursing, 13*(1), 51–58.

Unschuld, P. U. (2003). *Huang Di Nei Jing Su Wen: Nature, knowledge, imagery in an ancient Chinese medical text: With an appendix: The doctrine of the five periods and six Qi in the Huang Di Nei Jing Su Wen*. Berkeley: University of California Press.

Upenieks, V. (2003a). The interrelationship of organizational characteristics of magnet hospitals, nursing leadership and nursing job satisfaction. *The Health Care Manager, 22*(2), 83–98.

Upenieks, V. (2003b). What constitutes effective leadership? Perceptions of magnet and nonmagnet nurse leaders. *Journal of Nursing Administration, 33*(9), 456–467.

Valencius, C. B. (2000). Histories of medical geography. *Medical History Supplement, 20*, 3.

Valentine Seaman, (1798). The Medical Repository of Original Essays and Intelligence, Relative to Physic, Surgery, Chemistry, and Natural History (1797-1800), 1(3), 315.

Van Dongen, C. J. (2002). Environmental health and nursing practice: A survey of registered nurses. *Applied Nursing Research, 15*(2), 67–73.

Vanderheide, R., Moss, C., & Lee, S. (2013). Understanding moral habitability: A framework to enhance the quality of the clinical environment as a workplace. *Contemporary Nurse, 45*(1), 101–113.

Vessey, J. A., DeMarco, R. F., Gaffney, D. A., & Budin, W. C. (2009). Bullying of staff registered nurses in the workplace: A preliminary study for developing personal and organizational strategies for the transformation of hostile to healthy workplace environments. *Journal of Professional Nursing, 25*(5), 299–306.

von Strauss, E., Paillard-Borg, S., Holmgren, J., & Saaristo, P. (2017). Global nursing in an Ebola viral haemorrhagic fever outbreak: Before, during and after deployment. *Global Health Action, 10*(1), 1371427.

Watson, J. (1979). *Nursing: The philosophy and science of caring*. Denver: Colorado University Press.

Watterson, A., Thomson, P., Malcolm, C., Shepherd, A., & McIntosh, C. (2005). Integrating environmental health into nursing and midwifery practice. *Journal of Advanced Nursing, 49*(6), 665–674.

Watty, K., White, M., Matthews, J., Buchanan, L., Clarke, J., Sulkowski, A., et al. (2003). There's no place like home: A prospective evaluation of chemotherapy in the home. *Australian Journal of Cancer Nursing, 4*, 18–20.

West, C., Palmer, L., & Tier, T. (2000). No place like home: An antepartum program at British Columbia's Women's Hospital has shown that women with complications of pregnancy do better and save the system money when they are cared for at home. *Canadian Nurse, 96*, 32–34.

Whittaker, A., & Chee, H. L. (2015). Perceptions of an 'international hospital' in Thailand by medical travel patients: Cross-cultural tensions in a transnational space. *Social Science & Medicine, 124*, 290–297.

Williams, M. (2001). Critical care unit design: A nursing perspective. *Critical Care Nursing Quarterly, 24*(3), 35–42.

Williams, S. D., Hansen, K., Smithey, M., Burnley, J., Koplitz, M., Koyama, K., et al. (2014). Using social determinants of health to link health workforce diversity, care quality and access, and health disparities to achieve health equity in nursing. *Public Health Reports, 129*(1_Suppl 2), 32–36.

Wilson, L., Mendes, I. A. C., Klopper, H., Catrambone, C., Al-Maaitah, R., Norton, M. E., & Hill, M. (2016). 'Global health' and 'global nursing': Proposed definitions from The Global Advisory Panel on the Future of Nursing. *Journal of Advanced Nursing, 72*(7), 1529–1540.

Wong, F. K. Y., Liu, H., Wang, H., Anderson, D., Seib, C., & Molasiotis, A. (2015). Global nursing issues and development: Analysis of World Health Organization documents. *Journal of Nursing Scholarship, 47*(6), 574–583.

World Health Organization. (2016). *Global strategic directions for strengthening nursing and midwifery 2016-2020*. Geneva: WHO.

Wrede, S. (2010). Nursing: Globalization of a female-gendered profession. In *The Palgrave handbook of gender and healthcare* (pp. 437–453). London: Palgrave Macmillan.

Wright, D. J. (2003). Collaborative learning experiences for nursing students in environmental health. *Nursing Education Perspectives, 24*(4), 189–191.

Yost, J., Ganann, R., Thompson, D., Aloweni, F., Newman, K., Hazzan, A., et al. (2015). The effectiveness of knowledge translation interventions for promoting evidence-informed decision-making among nurses in tertiary care: A systematic review and meta-analysis. *Implementation Science, 10*(1), 98.

Chapter 2
The Geographical Turn in Contemporary Health Professional Research: Contexts, Motivators, Current and Emerging Perspectives

Gavin J. Andrews, Emma Rowland, and Elizabeth Peter

Abstract This chapter reviews the contemporary geographical study of health care work. Initially it examines key 'on the ground' transformations in health and health care which are fundamentally geographical in their making, form and consequences; transformations which have demanded a geographical research perspective be taken as well-aligned vantage point with which to report and understand them. These include increasing spatial diffusion of health professionals and roles; the transition of health care settings; the role of technologies in overcoming spatial limitations; the increasing emphasis on community and the social model of health; the embeddedness of geographical scales and concepts in policy and administration, and the globalization of work roles and responsibilities. The chapter then moves on to examine how a number of academic developments—including in medical/health geography—have at the same time coalesced to provide an additional set of motivations and opportunities for geographical scholarship on health care work. Finally, it describes the main theoretical traditions and approaches which have constituted 'geographies of health care work' as a current multidisciplinary academic enterprise and field—including spatial science, political economy, social constructionist/ humanist, and most recently non-representational theory, providing examples of recent empirical research which has been framed by each.

Introduction

This chapter provides a comprehensive review of the contemporary geographical study of professional health care work. Initially, it examines three recent 'on the ground' transformations in society and health care, which are fundamentally geographical in their making, form and consequences; hence transformations which have demanded an aligned geographical research perspective to report and understand them and their impacts. It then moves on to examine how a number of aca-

G. J. Andrews et al., *Place and Professional Practice*, Global Perspectives on Health Geography, https://doi.org/10.1007/978-3-030-64179-5_2

demic disciplinary developments have at the same time coalesced to provide an additional set of sectorial motivations and opportunities for geographical scholarship on professional health care work. Finally, it describes the main theoretical traditions and approaches which have constituted 'geographies of health care work' as a multidisciplinary academic enterprise and field, providing examples of recent empirical research which has been framed by each.

Three 'On the Ground' Geographical Transformations as Motivators for Geographical Health Professional Research

The 'Posthuman' Social Condition: Its Geographical Forms and Consequences for Health and Health Care

The first change described is certainly the most complex, but we argue it is most fundamental and underlies all others. Indeed, a very important development is an emerging 'posthuman' social condition that transforms all environments humans frequent, including spaces of health and health care. Moving beyond early popular notions of cyborg bodies and robot futures, the idea of posthuman is based on the observation that in the twenty-first century the idea of the autonomous, complete, fully conscious 'human subject' no longer accurately reflects the conditions of lived experience, and that more generally humans should not be taken as the measure of all things. Instead, humans are increasingly dependent, incomplete often less-than-fully conscious actors (Andrews 2019; Andrews and Duff 2019a). Under unprecedented technological advancement, humans increasingly operate in environments heavily dictated by digital cultures, algorithmic automation and engineering solutions. It is argued that five facets of the posthuman social condition are particularly prominent (see Andrews 2019; Andrews and Duff 2019a):

The first is the direct and increasing influence of hardware and software on people's lives. Hardware and software certainly give humans new forms of knowledge and awareness on topics including health, yet they can become obsessions in themselves through the alluring synthetic aesthetics and textures they provide, and affective forces they produce, that play to multiple senses and emotions. In particular, occupied by people for ever increasing amounts of their working and private time, cyberspace circumvents physical space and distance, with positive and negative health consequences (Marchant and O'Donohoe 2019; Tyrawski and DeAndrea 2015). Ultimately, looking to the not too distant future, artificial and super intelligence has the potential to extend human capabilities but challenges human primacy, concerns rightfully existing for insuring safeguards and limits and human wellbeing.

The second facet of the posthuman condition is an ever-dense materiality in consumer culture. This is produced by technologies, involves technologies as products, is circulated and transacted by technologies (Andrews 2019). All human cohorts are consistently targeted by, and exposed to, commercial affective forces in the environments they frequent. Whilst consumer agency can empower and provide enjoyment on many levels, some people are harmed by related obsessions, are excluded from it or are threatened by its externalities—such as pollution—that result from local to global scales (Andrews 2019). On the whole, without dramatic adjustment, this is not a sustainable, healthy future for humans or the planet.

The third facet is the increasing surveillance and monitoring of public and private lives and space. Whilst surveillance and monitoring technologies can be used for the good—such as for public and personal safety, fitness, nutrition, medication compliance, independent and lower risk living—often they provide personal information to other parties to be used either for commercial ends or to inform institutional responses (Andrews 2019; Andrews and Duff 2019a). In the end, surveillance both assists and detracts from health. Ultimately, it changes what it is to be a member of society, and to live and work in public and private space.

The fourth facet is new forms of life sustaining, life extending and life enhancing interventions and technologies. Highly varied, these are applied broadly in the realms of cosmetic and medical surgeries and drugs and supplements. They involve the insertion of materials onto and into the human body, or the material manipulation of the human body, and represent the latest iteration of the long-standing 'scientific' objective of a technologically-enhanced human body transcending natural limits (Andrews 2019; Andrews and Duff 2019a, b). This futurist goal is particularly prominent in the bio-sciences where the limits of the human body is regarded as a 'problem' to be overcome through technological means. Of course such a goal undoubtedly assists humans and humanity, yet can be selective, inequitable and ever supersedes the natural human. It certainly creates ethical, practice and economic debates on the types and extent of efforts society should exert in this direction and the types of humans that will emerge and need to be cared for in future.

The fifth facet of the posthuman condition is the proliferation of information. We have already mentioned this in the context of surveillance. More generally, it is clear that more information can personally inform and empower but can also lead to greater anxiety (through, for example, the way the 24-h news cycle presents it). An important development has been the emergence, through varied technological extraction techniques, of 'big data'—often on health. Big data might be collected by either the public or private sectors, drawn from media, web and other non-traditional sources as well as traditional ones (Andrews and Duff 2019b). This is a situation that promises new speeds and technologies of analysis and knowledge translation on the challenges of health and rapid responsive administrative/managerial decision making and policy making. Yet it is a situation that needs to be critically appraised and approached in terms of its powerful downstream transformative potential on spaces of health and health care.

Changing Geographical Dimensions to Health Professional Roles: Three Scales of Impact

Looking within the posthuman condition at some specific sectorial changes, it is clear that health professional roles and have altered dramatically. At the macro-scale through globalization whereby a wide-range of phenomenon—including policy, employers, representative organizations, regulators, evidence, finance, workforce and responsibilities—now reach across vast distances and bridge multiple territorial jurisdictions. Indeed, health professionals are increasingly working in a global environment, and are themselves increasingly globally informed (Falk-Rafael 2006; Herdman 2004; Jones and Sherwood 2014). In the twenty-first century, one can no longer think about geographically discrete pockets of practice located around the world. One must instead understand practice as a complexly inter-scaled networked global phenomenon.

At the meso-scale recent decades have witnessed the increasing spatial diffusion of professional roles. Whereas for many post-war decades almost all professional health care work was hospital-based—hence tightly spatially bounded—things have changed dramatically. Specifically, spatial diffusion has occurred in three ways: (1) through the growth in home-based professional support and care; (2) through growth in smaller specialized community-based settings/clinics for professional support and care; and (3) through hospitals providing outreach services that extend professional roles beyond their main sites. Indeed, in many respects professional practice is now found 'here, there and everywhere' where people live, work, travel, are educated, entertained and even shop (see McKeever and Coyte 2002). At the same time geographical change has involved an allied transfer of responsibility to communities. At one time viewed by policy makers and managers as administrative areas, communities have now been re-positioned as interconnected social terrains for health care, which of course have needs to be serviced but also the capacity to support and care for themselves as partners (Andrews and Evans 2008). Underpinning these trends are new technologies that facilitate 'remote' care and the development of new services through improving communication, monitoring and procedures. Contemporary examples include telehealth, telephone triage, remote monitoring, intervention and assistive technologies and distance diagnostic, acute and rehabilitative care (Lehoux 2008).

Meanwhile, within both hospitals and community settings, interpersonal changes have occurred at the micro-scale. The introduction of new technologies, new workload expectations and new role definitions have changed socio-psycho-physical relationships between people making, for example, health professionals more physically, emotionally and narratively distant from clients as their tasks become orientated around monitoring rather than building individual interpersonal relationships often at the bedside (Malone 2003). Admittedly, at a stretch, this change is somewhat consistent with past concerns for protecting clients' 'personal spaces' (i.e. comfort zones that could be intruded/violated, causing anxiety) and their 'interpersonal comfort' (a combination of their personal space whilst maintaining respectful appropriate conversation practice (Smith and Cantrell 1988). However, recent change arguably now goes much further than that, ultimately being to the detriment of humanistic therapeutic experience (Malone 2003).

The Prominence of 'Place' in Public Health, Health Policy, Facility Design Use and Representation

In recent years, the concept of 'place' has become increasingly popular in many changes and new priorities that affect health professional roles and work. The largest and most pervasive is a move from an exclusively medical model of health (scientific, biological, curative) to also a social model of health (multidisciplinary, multifaceted, preventative). This involves an acknowledgement that health and disease are influenced by place-based factors, and that where one lives, works, socializes and how one uses one's environment, has profound health implications (Andrews and Evans 2008). Indeed, this increasingly informing public health initiatives and practice the world over.

In other contexts, place has become an important concept in the framing of health policy and administration. Falling under a neo-liberal agenda, there have been two related developments. First is geographical decentralization and the move towards greater 'local control' in the running or supplementation of public health, health care and social care (also described under the terms 'regionalization', 'devolution' and 'deconcentration' (see Secker et al. 2006)). Second, set within the first, is the increased frequency of specific place-based initiatives. For example, area-based initiatives (Thomson 2008), settings-based approaches (Whitelaw et al. 2001) localised 'big society' initiatives (Taylor et al. 2011), age friendly communities (Plouffe and Kalache 2010) and 'aging-in-place' (Wiles et al. 2012). All these have important consequences for the types of work health professionals undertake and the responsibilities they have.

Finally, place comes to the fore, in particular political ideologies and directives involving the subjection of health care to market principles and conditions. As a result, in recent decades, health care places have changed not only in terms of their function but also in their character and how they are experienced. The modernist scientific hospital—long time a symbol of dull sterility, regularity and productivity—is being reinvented in many countries under corporate culture and new managerial modes as a postmodern consumer space that is inviting, exciting and even adventurous. Consumerist ideology in health care translates on the ground into what might be conceptualised as 'consumption landscapes' (Gesler and Kearns 2002), whereby either health care institutions, as whole entities, embrace market principles and/or the market colonizes parts of them (Kearns and Barnett 1997; Moon and Brown 1998). In terms of 'embracing' the market, it has been noted how hospitals increasingly deploy commercial language and self-promotional communication strategies aimed at a full audience of funders, donors, private business, government and the public. Here the plan is often to reduce negative associations between them and post-war institutionalized medicine, and to sell the new atmospherics and opportunities they offer whilst, at the same time, promoting their medical achievements and prowess. This being the case both in developed world contexts (Kearns et al. 2003) and increasingly in the developing world particularly with regard to medical tourism (Solomon 2011). In terms of the market 'colonizing', health care spaces are increasingly being opened up to commercial design and enterprise to make them more exciting, appealing and ultimately revenue generating (for example the locating of art, eateries, play spaces and retail spaces in hospital atriums (Adams et al. 2010; Kearns and Barnett 2000). This is not to say, however, that only large hospitals have become consumption landscapes, and ample evidence exists for how small community-based facilities of various types are now designed with atmospherics in mind, opened up to the market and promoted in the same ways (Andrews et al. 2013). In sum, new consumption landscapes raise questions concerning the embodied emotional responses of not only clients, families and other visitors, but also of health professionals and the changing nature of how their workplaces and labour are regarded.

Four Academic Developments as Motivators for Geographical Health Professional Research

Social Science and Humanities Perspectives in the Maturing of Health Professional Research

The introduction of a geographical lens in health professional research might be considered part of the natural evolution and development of the field, which over the past few decades has developed a range of social science and humanities lenses to help unpack and inform the intricacies of professional roles and contexts including social psychology (Paley 2014; Sah and Fugh-Berman 2013; Whitehead 2001), sociology (Aranda and Law 2007; Jacob 2017; Porter and Ryan 1996), anthropology (DeSantis 1994; Holden and Littlewood 2015; Holland 1999; Ovesen and Trankell 2010), political science (Browne 2001; Mason et al. 2013), economics (Aiken 2008; Croxson et al. 2001; Morris 1998), music and other arts (Jensen and Curtis 2008; McCaffrey and Locsin 2002), religious studies (Mickley et al. 1995; Papazisis et al. 2014; Sloan et al. 1999), history (Dingwall et al. 2002; Judd and Sitzman 2014; Scull 2015), ethics (Bandman and Bandman 2002; Butts and Rich 2019; Liaschenko and Peter 2004), and social theory and philosophy (Cody 2006; Edwards 1997; Van der Eijk 2005).

Social science and humanities perspectives are important in understanding health care work because they reflect 'on the ground' practice realities. For example, to some extent sociology has proved insightful to professional practice because health care does not exist in a vacuum, and the social nature of it and its contexts are critical. By the same token, social psychology has proved insightful to professional practice because health care workers, their clients and the public constantly ruminate on, and make assessments on, health matters. Economics has proved relevant to professional practice because workers increasingly take responsibility for, or are themselves, scarce resources (and so on). Health professionals certainly do not to think of these as strange or controversial perspectives to take and disciplines to draw on. Hence geography, with its particular conceptual claim that space and place matter to practice, is simply one of a number of useful lenses health professional researchers have adopted to help better know their roles, clients and environments and shape their own agendas (Rafferty 1995). Admittedly, the infusion of social science and humanities perspectives in health professional research over the years has not been linear. Certain places/countries have led the whilst others have lagged, and there has certainly existed a history of growth and retraction, both largely as a result of changing priorities and funding in health professional education and research (Kyle et al. 2016).

Geographical/Spatial Turns Occurring Across Academia

The introduction of geographical perspectives in contemporary health professional research is also part of a broader 'spatial turn' that, in the last two decades, has occurred across a range of health, social science and humanities academic disciplines. At the heart of this turn is an acknowledgement that a wide range of spatial phenomena matter (including distributions, locations and distances, urbanicity, rurality, settings, landscape, mobility, sense of place and atmosphere). Part of the impetus for this spatial turn has been technical, involving the increasing obtainability of technology and data suited to undertake spatial representation and analysis (particularly with regard to Geographical Information Systems—GIS), whilst part has been theoretical; a realisation that personal, social, cultural and economic life are each complexly 'emplaced' (Bergmann 2007; Richardson et al. 2013; Van Ingen 2003; Warf and Arias 2008). With regard to the latter, particularly influential has been the work of social theorists whose ideas are explicitly spatial. Most notably, this has included the work of Heidegger (embedded knowledge), Simmel (social geometry), Goffman (institutional space), Foucault (the gaze, surveillance), Said (orientalism), Lefebvre (the production of space), Harvey (time-space compression and space-economy), Bourdieu (habitus), Deleuze (affect), Soja (thirdspace) and Merleau-Ponty (spatiality) (e.g. Doona et al. 1997; Gastaldo and Holmes 1999; Nairn and Pinnock 2017). Health professional researchers have often used these theorists' ideas to frame their studies that then have ended up asking relatively spatial research questions and/or reporting relatively spatial findings as a result. In addition, important to the spatial turn has been the continued growth and prosperity of spatial disciplines other than geography including environmental psychology, environmental health, architecture and planning, landscape epidemiology and population health (e.g. Keller et al. 2004; Timmermans et al. 1998).

The Gap Created by Health Geography's 'Post-Medical' Neglect of Health Care Work

Health geography has certainly neglected the topic of professional health care work on various levels and scales (Andrews and Evans 2008; Connell and Walton-Roberts 2016), this neglect arguably creating an opportunity and need for health professional disciplines to fill the resulting gap. Of course, there are exceptions (Batnitzky and McDowell 2011; Castleden et al. 2010; Rapport et al. 2007; Yantzi and Skinner 2009). Other than the current book editors' work (Andrews and Shaw 2008; Andrews and Moon 2005a, b), most notably for example being critical work on labour supply by Kim England (England 2015; England and Henry 2013), Margaret Walton-Roberts (Walton-Roberts et al. 2017a, b; Walton-Roberts 2012, 2015, 2019, 2020; Thompson and Walton-Roberts 2019; Ennis and Walton-Roberts 2018) and John Connell (Brown and Connell 2004; Connell 2007, 2008, 2010a, b, 2014; Connell et al. 2007). Moreover, also notable is theoretical/positional work on nursing by Richard Kyle and Iain Atherton discussed later (Kyle and Atherton 2014, 2016; Kyle et al. 2016).

Exactly why health geography has neglected professional health care work a complex question. One of the likely reasons is caught up with a sub-disciplinary change in the 1990s from 'medical geography' (the study of disease and health care) to 'health geography' (including the study of health, wellness and wellbeing). Arguably, as part of the politics of this change, a 'post-medical' movement or at least sentiment arose, which aimed to take the sub-discipline out from the shadow of medicine. Hence for some geographers—at least for a time—'post-medical geography' evolved into a catch all sub-disciplinary phrase, which negatively emphasized, and dismissed, what is no longer a priority (e.g. Kearns 1993; Parr 1998a, b). In terms of meaning, post-medical geography provided a proud non-medical identity to the subdiscipline. Unfortunately, however, as we shall describe, post-medical geography legitimized a rather narrow view of ideal sub-disciplinary progress and created missed opportunities in research.

Numerous problems have been highlighted with the idea of post-medical, but two are most prominent. (1) Critics have argued that one problem is that any 'post' prefix is understood to mean the death of that which follows it (for example post-mortem). Moreover, it fixes the meaning of its root and any use becomes dependent upon a consensual understanding of this fixing (for example, the term postmodernism has been undermined because some scholars have persuasively argued that society has never been truly modern). The same types of problems occur with post-medical geography. Medical concerns still arise in so many contexts related to health, and cannot easily be separated or ignored. (2) A further problem regards the nature of medicine. The term post-medical can be thought of as part of a reactionary rhetoric that promotes a simplistic and dated caricature of medicine as something universal, universally powerful and even uncaring. This artificially sets up medicine as a convenient 'folk devil' to be progressively distanced and ignores medicine's diversity, sensitivity and capacity for self-reflection and change (Kearns and Moon 2002).

Having said all this, it is still a little surprising that health geography has neglected professional health care work as a subject given the relatively strong record that it and human geography more generally has in studying other forms of work. Social and health geography, for example, together have forged at least three relevant fields of study. First, the intricacies of 'allied' professional work has been an interest, including focuses on forms of holistic medicine, health promotion, management administrative and NGO roles (Andrews and Phillips 2002; Bauder 2003; Doel and Segrott 2004; England 2010; Hanlon 2001; Lovell et al. 2011; Neuwelt et al. 2015; O'Reilly 2011). Second, informal/voluntary forms of care and welfare work has been a concern much of this focus on specialist groups (Conradson 2003; Herron and Skinner 2012; Milligan 2017; Power 2008; Skinner 2014; Wiles 2003). Third, wellbeing in everyday work has been of interest to social and health geography including focuses on balancing domestic and paid workspaces, working *with* personal disabilities and illness and coping *in* mobile and/or marginalised community-based work (Cahuas 2019; Dyck 1989, 1990; Dyck 1999; Evans and Wilton 2019; Gorman-Murray and Bissell 2018; Hall 1999).

Beyond this, a 'new' economic geography has in recent years provided a revitalised and critical perspective on work. At the macro-scale it has broadly considered labour organization, transformation and relations (Castree et al. 2004a, b; Hastings 2016; Hudson 2001). At the meso and micro-scales, it has considered various intricacies, interpersonal and performative aspects of work, workspaces and working life (James 2017; Wills et al. 2000; Worth 2016). Here, interest has ranged from small to large businesses and has included sectors such as retail (Crew et al. 2003; Everts 2010; Gregson et al. 2002), hospitality (Crang 1994; Guyatt 2005; McDowell et al. 2007), arts (Cameron 2007; Watson and Ward 2013), elite business (Cormode and Hughes 1999; McDowell and Court 1994), training, knowledge and information (Cranston 2014; Hastings and MacKinnon 2017; Pykett and Enright 2016) and hi-tech and IT (Blumen 2012; James 2011; Pottie-Sherman and Lynch 2019). Meanwhile, an occasional interest here is public and private sector manual work, often placed in particular historical socio-political and socio-economic contexts (Gregson et al. 2016; Kobayashi 1994; Rogaly and Qureshi 2017). These lenses and expertise—what might be thought on as 'critical labor studies'—and particularly their perspectives on power, surveillance and emotions in work, potentially have much to offer and could easily have been be brought to bear in the study of professional health care. Indeed, we do exactly this in Chaps. 3–5.

Key Academic Initiatives with a Geographical Emphasis

Although extremely varied, a number of academic initiatives have also acted as nodes and stimuli for the expansion of geographical health professional research; informing education, focusing scholars, research inquiries and increasing research capacity. An important example is the *In Sickness and in Health* bi-annual international nursing conference series, which since 2002 has brought together and showcased a large volume of critical scholarship. Although not explicitly geographical, key research headed by or involving many of the original founding scholars specifically unpacks power in professional practices and places (e.g. Björnsdóttir et al. 2015; Barnes and Rudge 2005; Ceci 2004; Gastaldo et al. 2004; Holmes 2001; Mapedzahama et al. 2012; Nelson et al. 2011; Pryce 2004; Purkis 2001; Racine and Perron 2012). Another example is the Canadian Institutes of Health Research (CIHR) funded strategic research and training program *Health Care Technology and Place* (HCTP) launched in 2003 by Peter Coyte and Patricia McKeever, based in Canada but networked internationally. This group brought health professionals, health scientists and social scientists together over a 12-year period until 2015, conceptualizing the many ways in which 'place matters' in twenty-first century health care, resulting in numerous collaborative projects and publications (e.g. Coyte and Holmes 2006, 2007; Crooks and Agarwal 2008; Holmes et al. 2007a, b; Lehoux et al. 2007, 2008; King et al. 2018a, b; Miller et al. 2003; Poland et al. 2005). Most recently, continuing the tradition of team initiatives is the British, Economic and Social Research Council (ESRC) funded *Social Science and Nursing* project that ran between 2014 and 2016 (involving Richard Kyle, Iain Atherton, Sally Haw, Kate Seers, Stephen Timmons and Glenn Robert). It considered how social sciences can be incorporated to a greater degree into nursing education and practice; geography being given particular prominence in key seminars and publications over the 3-year period (Atherton and Kyle 2015; Kyle and Atherton 2014, 2016; Kyle et al. 2016). Other notable initiatives include forms of collectives/associations that act as nodes to focus support, education, research and practice for specific types of geographical issues including rural and remote nursing (carrn.com), nursing and the environment (cnhe-iise.ca and envirn.org), nursing in indigenous communities (indigenousnurses.ca) and street nursing (streethealth.ca).

In terms of specific publishing endeavours, notable has been a ground-breaking special issue of the *International Journal of Older People Nursing* on the importance of place in older person's care. These papers providing key theoretical reviews (Andrews et al. 2005a, b; Cutchin 2005; Wiles 2005). Likewise, important was a 2007 special issue of *The Canadian Journal of Nursing Research* focused on 'the importance of geography and health in nursing research'. Its seven empirical papers explore the varied places where nursing occurs and where the people and populations they care for reside, and associated issues and challenges (see Bender et al. 2007; Crooks et al. 2007; Etowa et al. 2007; Hodgins and Wuest 2007; Rutakumwa and Krogman 2007; Solberg and Way 2007; Tarlier et al. 2007). Meanwhile, of positive influence has been the various editors, editorial boards and reviewers that over recent years have welcomed geographical health professional research into other journals, most notably *Health and Place, Social Science and Medicine, Nursing Inquiry, Nursing Philosophy, Journal of Advanced Nursing, and Aporia.*

Theoretical Traditions in Geographical Health Professional Research

Geographical perspectives and research can be traced across a number of professional categories, most notably nursing (Andrews 2006) and general medicine (Groutsis 2006), but also occupational therapy (Rowles 2000), pharmacy (Thompson and Bidwell 2015), public health (Dummer 2008), family practice (Agarwal 2009), dentistry (Levin et al. 2010), complementary medicine (Anyinam 1990), physiotherapy and rehab (Sandal et al. 2019), psychiatry (Fullilove 1996), laboratory science (Greenhough 2006), midwifery (Hammond et al. 2013) and social work (Hudson 2012). In addition, there also is a strong interprofessional interest (Brewer et al. 2017; Poland et al. 2005). Across these categories, and others besides, a number of theoretical perspectives have emerged:

Spatial Science and Political Economy

A positivistic 'spatial science' tradition has strongly emerged, concerned with aerial differentiation; the quantitative calculation and possible mapping of distributive trends across space (i.e. people, diseases, services and other resources) often using Geographical Information Science (GIS) as an analytical and representational tool (for example Bickes 2000; Blake and Bentov 2001; Bloch 2011; Courtney 2005; Faruque et al. 2003; Endacott et al. 2009; Graves 2012a, b; Gesler et al. 2004; Lin et al. 1997; Lee and Park 2006; Moss and Schell 2004). Indeed, this project has involved the use of deterministic models, inferential statistics, probability testing and other approaches to find spatial patterns in health and health care phenomenon important to health care work or of work itself. Underpinning this research is a particular conceptualization of space: absolute space—a featureless, neutral surface upon which life unfolds. Interesting things happening in research when data points (such as related to people, disease or facilities) are located in space. Space then becomes mathematically distinguishable and begins to factor as a substantial feature and challenge. *At* these locations, rates, volumes and other localized measures become visible (such as levels of morbidity and mortality and various metrics for service performance). *Between* these locations, times, distances, movements and differences became visible (such as the spread of disease, expansion of services, travel to facilities).

Arising in many cases simultaneously with spatial science has been a political economy tradition in geographical health professional research. This is based on a neo-Marxist understanding that spatial arrangements of resources are aligned closely with labour and economic relations. Specifically in terms of empirical research, it is concerned then with how health systems and policies play out spatially and the relationship to health needs and outcomes. Notably, like with spatial science, political economy is often motivated by ideas around 'distributive justice' and searches for the most optimal/ethical spatial allocations of resources (for example Aiken et al. 2004; Brush and Sochalski 2007; Kingma 2003, 2006). Indeed, in explaining distributive features over international, national, regional and localized areas, this is a theoretical tradition that contributes directly to mainstream health service and academic debates on rationing, efficiency and equity in service, planning and provision (Joseph and Phillips 1984).

Empirically, there has been some limited interest in geographical issues at a micro-scale. Studies conducting, for example, classical ‘time and motion’ studies of work (Burke et al. 2000; Hendrich et al. 2008, 2009). However, most attention has been to contextual meso and macro-scale geographical issues. On the one hand, there has been a long-term consideration of the distributive features and concerns of particular client and population groups health professionals have responsibility for (see Gesler et al. 2004; Hodgins and Wuest 2007; Moss and Schell 2004; Thomas 2013a), most recently in the context of medical tourism and the movements of clients around the world (Crush and Chikanda 2015; De Arellano 2007; Horowitz et al. 2007; Horowitz 2007; Johnston et al. 2010; Lunt and Carrera 2010; York 2008). On the other hand, a prime concern is distributive features of health care workforces, a longstanding specialist field of study here being for example paramedic work, and issues such as optimal allocations, routes and response times (Abbott 2008; Beillon et al. 2009; Mayer 1979; McMeekin et al. 2014; Ong et al. 2009; Peleg and Pliskin 2004; Uyeno and Seeberg 1984; Williams and Shavlik 1979). By far most research on workforces, however, is on career movements. Attention is paid to experiential aspects and the social, political and economic forces that shape them at local (Brodie et al. 2005), national (Cho et al. 2014; Courtney 2005; Harris et al. 2013; Kovner et al. 2011; Lin et al. 1997; McGillis Hall et al. 2013; Radcliffe 1999) and international scales (Aiken et al. 2004; Bach 2015; Brush and Sochalski 2007; Buchan et al. 1997; Buchan 2001; Buchan and Sochalski 2004; George 2015; Kingma 2001, 2006, 2007, 2008; Kline 2003; McGillis Hall et al. 2009; Ross et al. 2005).

Meanwhile, in another field of study decision-making (as very broadly defined) has also been subjected to a political economy lens in two areas. On the one hand, in studies of the geographical dimensions to, or consequences of, decision making across cohorts of decision-makers (for example family or hospital-based doctors or service managers). These decisions are often quite defined in nature, involving specific financial, planning or clinical concerns and are often reactions to broad policy and/or system incentives or changes (see Carr-Hill et al. 1994; Iredale et al. 2005; Moon et al. 2002; Twigg 1999). On the other hand, very much related to the aforementioned literature on the movements of workforce, other studies focus specifically on locational career decision-making processes, and the consequences of these decisions for the supply of labour, service viability and quality and local communities (Baer 2003; Barnett 1988, 1991; Cutchin 1997; Farmer et al. 2003; Guagliardo et al. 2004; Harrison 1995; Kazanjian and Pagliccia 1996; Laditka 2004; Wei 2004).

As is clear from these areas of interest, spatial science and political economy are mostly about where resources and challenges lie. This literature is particularly important as it relates to planning, yet is has been left to more humanistic research to get to grips with the finer details of practice and what health professionals do in care in their everyday working lives.

Humanism and Social Constructionism

A tradition that draws on humanistic and social constructionist theory is concerned with how practices and places (in this case clinical practice and settings) represent and make people (in this case health professionals and clients), and vice-versa how people represent and make practices and places through ongoing processes of meaning-making (e.g. Halford and Leonard 2003; Cheek 2004; Andes and Shattell 2006; Kitto et al. 2013; Liu et al. 2014; George 2015). A look at how this works out theoretically is worthwhile. Following precedent in human geography and other social science disciplines (Kearns 1993), an understanding has developed that, at one level, because of human roles and structural features in situ, places possess basic agency (hospitals provide institutional medicine, community clinics provide primary health care and so on). At another level, however, underlying this basic agency, intimate experiential processes are recognized to be at work. These start with the idea of 'imbedded knowledge', which, based on Heidegger's thinking, posits that humans can only relate to and beyond themselves through their situation, their literally 'being-in-the-world' and their relational consciousness of other humans and non-humans in the world (Bender et al. 2010; Andrews 2016). It follows then that 'imbedded knowledge' can be gained through two forms of encounter with place. First, through direct encounter, and the fundamental human condition of being bodily and mentally present in place. Second, through partial encounter; this this not involving physical presence but knowledge attained through other means such as internet, media, art, literature or word-of-mouth. Both forms of encounter allow, in Husserl(ian) terms, the 'intentionality' and 'essences' of places to emerge (Andrews 2016). With respect to 'intentionality', it is posited that through human presence, perception and judgment, places are ascribed meaning. From a phenomenological standpoint, just as the use of an object is critical to its meaning (i.e. the object is 'about' what humans do *with* it) so are the uses of a places critical to its meaning (i.e. a places is 'about' what humans do *in* it) (Bender et al. 2010). With regard to 'essences', just as an object possess essences (i.e. its facets that influence what humans feel emotionally *about* it), so do places (i.e. its qualities that influence what humans feel emotionally *about* and *in* it) (Bender et al. 2010). Indeed, this is how a 'sense of place' emerges, how places can evoke a broad range of emotions, ranging from very positive to very negative (Relph 1976; Tuan 1977; Eyles and Williams 2008). The argument follows that such feelings, when experienced repeatedly or over time, can develop identities and attachments to places. Moreover, in terms of collective qualities, when members of particular groups (such as health professionals or members of a local community) share identities with, attachments to and agency in places, unique 'cultures of places' develop (such as specific workplace cultures or neighborhood cultures) (Andrews 2016).

In health professional research, these kinds of theoretical understandings have underpinned what might be described as a 'landscapes as texts' approach, whereby places are understood and approached in the same way as written material—as phenomenon whose structures and agencies can be, largely through qualitative methods, read, decoded, comprehended and re-written. Empirically, there has been a focus on clients, such as from critical human rights and post-colonial perspectives, focused on the places, situations and experiences of international migrants (Gastaldo and Magalhaes 2010; Seto-Nielsen et al. 2013). Most research, however, unpacks how specific professional career categories and job types and their practices relate to place (see Andrews 2006; Carolan et al. 2006; Stuart et al. 2008; McIntosh 2008; Thomas 2013b; Atherton and Kyle 2014). A range of relationships has been articulated, including how places are attributed symbolism and identity by, and in relation to, health care professionals but are oftentimes disputed and contested (Savage 1997; Halford and Leonard 2003, 2006a, b; Cheek 2004; Gilmour 2006), and thus how places typify and express particular professional disciplines (and vice-versa) (Roush and Cox 2000; Andrews et al. 2005a; Thompson 2008; Shattell et al. 2008; Oandasan et al. 2009; Lapum et al. 2009; Davis and Walker 2010; Corrêa et al. 2013). For example, place has recently been articulated as critical to the nature of community-based practice (Bender et al. 2007), home care (Duke and Street 2003), mental health care (Montgomery 2001; Andes and Shattell 2006), midwifery (Lock and Gibb 2003; Watson et al. 2007), pediatrics (Zitzelsberger et al. 2014) and gerontology (Cheek 2004; Wiles 2005).

Meanwhile, further studies have investigated the two-way dynamics between places and four domains. (1) Worker–patient decisions, ethics, interactions, emotions and relationships (Liaschenko 1994, 1996a, b, 1997, 2000, 2001, 2003; Purkis 1996; Peter 2002, 2003; Malone 2003; Bucknall 2003; Peter and Liaschenko 2004; Andrews and Peter 2006; Shattell et al. 2008; Andrews and Shaw 2010, 2012; Liaschenko et al. 2011; Seto-Nielsen et al. 2013). (2) Intra and inter-professional exchanges and relationships (Sandelowski 2002; Halford and Leonard 2003, 2006a, b; West and Barron 2005; Barnes and Rudge 2005; Brodie et al. 2005; Oandasan et al. 2009; Mesman 2012; Kitto et al. 2013). (3) The nature and outcomes of care, including specific types of clinical and caring interventions (McKeever et al. 2002; Angus et al. 2003; Hodnett et al. 2005, 2009; Marshall 2008; Mesman 2009, 2012). (4) Community practice and cared-for communities (Hall 1996; Pardo Mora and González Ballesteros 2007; Bender et al. 2007). Here, the focus had included the nature of rurality (Bigbee 1993; Shreffler 1996; Leipert and Reutter 1998; Skelly et al. 2002; Bushy 2002, 2008; Tarlier et al. 2007; Thomas 2011; Leipert and Anderson 2012) and urbanicity (Affonso et al. 2004; Vandemark 2007; Skott and Lundgren 2009; Bender et al. 2010; DeGuzman and Kulbok 2012; Thomas 2013a).

Humanistic and social constructionist studies are not, however, only focused on contemporary contexts, and go to the very origins of health professions themselves. For example, crosscutting many of the aforementioned themes related to emplacement and identity, 'historical geographies' illuminate alternative, often overlooked, histories of professional practice rooted in places that lie outside—and challenge—dominant western narratives. Indeed, this approach has been exemplified in relation to a wide-range of regions and countries (see Nelson 2001; Arseneault 2008; Burnett 2008; Daigle 2008; Elliott et al. 2008; Elliott 2008; McKay 2008; Quiney 2008; Rutherdale 2008; Stuart 2008; Toman 2008). Another specific example is critical research that examines the relations and politics of missionary medicine in past centuries (Kakar 1996; Good 2004; Jennings 2008; Kalusa 2007; Mohr 2009).

The two theoretical traditions covered in this section are, of course, not mutually exclusive and do not preclude a range of other, often overlapping, theoretical orientations that have emerged alongside them in geographical health professional research such as ecology (see Shreffler 1996; Huynh and Alderson 2009), feminism (see Liaschenko 1997; Peter 2002) and a wide-variety of poststructuralist theory (see Riley and Manias 2002; Andrews et al. 2005a; Thompson 2008). Notably, from these two traditions has emerged the human geography field of 'emotional geographies' which is used to frame the case studies in Chaps. 3 and 4.

The Uptake of Relational, Non-representational Theories

Most recently, geographical health professional research has begun to very tentatively explore new possibilities for inquiry. Underlying this has been adherence to a broad 'relational turn' in human geography and spatial health sciences that has arrived in the last 10 years (Cummins et al. 2007). This turn has criticized traditional understandings of place for being somewhat discrete and static. In response, the broad paradigm of relational thinking complicates some conventional assumptions. It, for example, introduces a twist in how place is theorized, evoking an image of places emerging not on their own, but through their connections within networks of other places (Jones 2009). Indeed, various aspects of geographical relational thinking are helpfully explained by Cummins et al. (2007), who emphasize that (1) a conventional view of place focuses on fixed boundaries, whereas a relational view focuses on fluid ever changing boundaries. (2) A conventional view of place focuses on their content, whereas a relational view focuses on influxes that change their content. (3) A conventional view of place focuses on residence whereas a relational view focuses on mobility between places. (4) A conventional view of place focuses on certain times and places, whereas a relational view focuses on many times and places and development through time. (5) A conventional view of place focuses on common understandings of them between individuals, whereas a relational view focuses on variable understandings of them between individuals.

Most recently, however, scholars have started to move beyond 'mapping' relationalities (i.e. articulating places as nodes within wider networks) and have embraced a further understanding that places are themselves relationally performed. In doing this they have started to edge close to 'non-representational theory', a broadly 'posthumanist' theory (posthumanism an overarching tradition itself examined in Chap. 6). In short, there are two parts to the 'non-representational' in non-representational theory. One is concerned with what is going on in the world. Rather than being concerned, as humanistic research is, with how and what humans contemplate and then purposefully communicate/represent (such as through language and text), it is instead concerned with less-than-fully conscious and physical modes of human action, expression and communication (Andrews 2017). The other part is concerned with how researchers might tell what is going on in the world. Rather than being concerned, as humanistic research is, with theorizing aspects of human life (i.e. digging for underlying explanations), it is instead concerned with 'showing' and 'animating' these aspects (i.e. letting them breathe and speak for themselves) (see Thrift 2008; Vannini 2009). With regard to 'theory' in non-representational theory, here it is meant as plural constituted of a number of 'contact zones' between a range of theoretical positions, including political and social ecology, neo-vitalism speculative realism, new materialisms, pragmatism and actor-network theory (Andrews 2017; Cadman 2009; Vannini 2009). Below we discuss non-representational theory's key facets (see also Andrews 2014, 2017, 2018). For each we mention some current health professional research which displays some close resonance.

The first of non-representational theory's facets is to recognize and show the '*whole onflow*' of life; the emerging frontier of existence, the leading point of the arrow of time. Based on a Deleuzian reading it is believed that life is initially expressed on a 'plane of immanence' a self-organising process through which it emerges (Deleuze and Guattari 1988). This plane—or virtual field—is subjectless, neutral and unconfined, preceding meaning or individualization, existing only through the singularity of events (Andrews 2018). Whole onflow is thus a pivot of sorts, which makes the world unpredictable, subject to immediate change and unlocks its unlimited creative potential. In terms of alignments in health professional research and practice, although scholars have yet to engage with whole onflow specifically, a number of literatures draw close. There has been (1) discussion of 'rhythm' in professional care (for example considering interactive patterns with patients (Newman 1999)); (2) discussion of 'momentum' in specific practices (for example considering distraction and continuance techniques nurses use when undertaking difficult procedures (Andrews and Shaw 2010, 2012)); (3) discussion of factors that interrupt or facilitate certain ambiances and flows in care (for example research on specific environmental factors—such as noise and design—that help or hinder flow (Cmiel et al. 2004; Hurst 2008; Morrison et al. 2004; Snyder 1973); (4) discussion of 'workflow' and how tasks follow on from each other smoothly or otherwise (Cornell et al. 2010a, b, 2011); and (5) discussion of time and motion in professional work (for example considering nurses' movements in medical wards (Hendrich et al. 2008, 2009)).

The second of non-representational theory's facets is a '*relational materialism*' in its approach. Non-representational theory sees the world as physically, complexly and unavoidably networked. Moreover, in these networks it recognizes the ontological equal importance and co-evolution of—and sometimes unclear distinction between—human and non-human entities. Importantly, it recognizes that relational materialities are performed (as human and nonhuman entities with relative positions and movements interact). With regard to alignments in health professional research, these relationalities are evident in a range of scholarship that considers the energy and agency that technologies and other objects possess and exert in practice contexts (Sandelowski 1996, 1999; Barnard and Sandelowski 2001; Wilson 2002; Zitzelsberger et al. 2014; Peter et al. 2014). This constituting part of a wider interest in technicity, materiality and actor-networks in health care work (Holmes et al. 2010; Hujala and Rissanen 2011; Lapum et al. 2012a; Einboden et al. 2013). Importantly non-representational theories recognize excessive outcomes of this relational materialism; vital forces resulting from vital living things acting in congress with vibrant objects lending life a self-generating impulse, continuance and purpose. This is reflected partially in health professional literature on holistic practice (Patterson 1998) and to some extend on the vitality, richness and good functioning of professionals care and organizations (Tucakovic 1999; Webster and Kristjanson 2002).

The third facet of non-representational theory is appreciating '*virtuality and multiplicity*' at play in the world. As Cadman (2009) suggests this has involved a Deleuzian reading whereby 'the virtual' is aligned with potential, but is something that is not always fulfilled in an actual space-time event, whilst 'multiplicity' refers to existence arising at many simultaneous times and levels, with simultaneous qualities from simultaneous viewpoints (a temporal heterogeneity whereby in the present several states permeate one another but combine into a rich conscious whole). Both speak straightforwardly to the way that space-time is ruptured—non-fixed and non-linear—and the relatedness of life across these ruptures (McHugh 2009). Current alignments with such thinking in health professional research include research on how technologies create new virtual cyber spaces for practice (Barnes and Rudge 2005; Mugavin 2014). Elsewhere in research on clients' mental dispositions, registers and thinking, and professional roles in encounters with these. Specifically, for example, in considerations of how visualization and imagination techniques create monetary mental 'spaces' that help relax and distract clients (Hoffart and Keene 1998; Andrews and Shaw 2010, 2012). Also in considerations of 'derealization' and 'disassociation' in mental health symptomology. Finally in considerations of hope, when no long-term cure exists, hope residing in the subtle adjustments that health professionals can make—such as buying into a patient's mental displacement and physically moving with them (Peter et al. 2014).

The fourth of non-representational theory's facets is a concern and eye for the '*processual*'. Non-representational theory recognises that life is a series of ongoing processes – actions that follow, and flow, with some order. In non-representational theory, there are three primary engagements with processual: with what happens (i.e. 'the processual *of* life'), how it happens (i.e. 'the processual *in* (processual) life') and how to investigate what happens (i.e. thinking processually to find and describe these two things). A vast swathe of health professional practice is, of course, processual, reflected for example in numerous research studies that are task-based, focused on the timing or ordering of events (Hendrich et al. 2008, 2009) and decision making from the individual to the organizational level. Particular alignment with non-representational theory is also evident in specialist literatures that consider the processual at transpersonal levels. For example, the personal process of 'self-reflective practice'; the relational process of 'transference' and the relational process of 'countertransference' (O'Kelly 1998; Jones 2004; Johns and Freshwater 2009). In each of these cases, a key consideration is how health professionals act into, continue, discontinue or divert the process bodily, individually, collectively, consciously and less-than-fully consciously.

The fifth facet of non-representational theory is *'practice and performance'*. A recognition being of a world that is industrious in every single moment, involving expressive arrangements and engagements of bodies and objects—geometries that make the world productive familiar and comprehensible (Thrift 2008; Andrews 2014, 2018). In this endeavor, the motivation is not to look for possible hidden signs and meanings that might be 'read', but instead to acknowledge core active elements for what they are including intervals, spacings and events (whether these be common or unique, intentional or coincidental, expected or unexpected, consciously undertaken and registered, or less-than-fully consciously undertaken and registered). In health professional research, partial precedent exists in basic time and motion studies, although these studies are often far more macro-scaled (focusing on cohorts of workers at the institutional level) and 'coarse' (undertaken over long time frames) (see Burke et al. 2000; Hendrich et al. 2008, 2009). Otherwise resonance exists in the study of health professional's postures, gestures, cadences, patterns and other performative aspects in space-time that make and reveal what they are (Raingruber 2001). Such inquiries inform, for example, debates on the nursing presence, and the specifically the idea and objective of 'culturally competent care'.

Non-representational theory's sixth facet is a concern to explore and animate *'transpersonal sensations and atmospheres', i.e.* sensations and atmospheres produced through, circulated amongst/traversing and shared between bodies. This particular facet has led to the uptake of the Deleuzian idea of 'affect' as an explanatory concept (see Thrift 2004a, b, 2008). Although affect has been subject of much debate, a general academic consensus is that it is an infectious process, in that it involves bodies affecting other bodies this resulting in increases or decreases in their individual and collective energy. Moreover, that affect is experiential; it being felt as the intensity of one's material/human environment and the intensity of one's involvement in it (Andrews et al. 2013). Somatically registered then as a vague yet powerful 'feeling state', affect is not a typical emotional category and experience (such as fear, love, loss, anger, empathy and numerous others). These are more personal, internal, specific and 'known', although affect can certainly underlie and/or co-exist with them (Pile 2010). A number of recent health professional studies consider the importance of affect in the making and experience of educational and clinical environments (see Ducey 2007, 2010; Rudge 2013; Solomon 2011). In particular, certain studies have considered the impacts of comfort, pain and other factors, which block or open up positive or negative affection in clients and influence their involvement environments (see Bissell 2008, 2009, 2010).

Summary

This chapter has reviewed a wide range of contexts, reasons and motivators behind the emergence of geographical health professional research, as well as the main and new theoretical traditions that have informed and framed it. With regard to the latter, we acknowledge that certain research draws on other theoretical traditions not described here, and that these traditions do not always align neatly in the ways presented. Nevertheless, our sections, we think, provide the most useful and appropriate general summary of perspectives. In the next chapters, we move on to present some in depth empirical research that showcases geographical health professional research.

References

Abbott, R. M. (2008). Geographic analysis of ambulance availability in Arkansas. *Family & Community Health, 31*(1), 35–43.

Adams, A., Theodore, D., Goldenberg, E., McLaren, C., & McKeever, P. (2010). Kids in the atrium: Comparing architectural intentions and children's experiences in a pediatric hospital lobby. *Social Science & Medicine, 70*(5), 658–667.

Affonso, D. D., Andrews, G. J., & Jeffs, L. (2004). The urban geography of SARS: Paradoxes and dilemmas in Toronto's health care. *Journal of Advanced Nursing, 45*(6), 568–578.

Agarwal, G. (2009). Geographies of family medicine: Describing the family doctor's practice-based landscape of care. In V. Crooks & G. J. Andrews (Eds.), *Primary health care: People, practice, place*. Milton Park: Routledge.

Aiken, L. H. (2008). Economics of nursing. *Policy, Politics, & Nursing Practice, 9*(2), 73–79.

Aiken, L. H., Buchan, J., Sochalski, J., Nichols, B., & Powell, M. (2004). Trends in international nurse migration. *Health Affairs, 23*(3), 69–77.

Andes, M., & Shattell, M. M. (2006). An exploration of the meanings of space and place in acute psychiatric care. *Issues in Mental Health Nursing, 27*(6), 699–707.

Andrews, G. J. (2006). Geographies of health in nursing. *Health & Place, 12*(1), 110.

Andrews, G. J. (2014). Co-creating health's lively, moving frontiers: Brief observations on the facets and possibilities of non-representational theory. *Health & Place, 30*, 165–170.

Andrews, G. J. (2016). Geographical thinking in nursing inquiry, part one: Locations, contents, meanings. *Nursing Philosophy, 17*(4), 262–281.

Andrews, G. J. (2017). Geographical thinking in nursing inquiry, part two: Performance, possibility, and non-representational theory. *Nursing Philosophy, 18*(2), e12137.

Andrews, G. J. (2018). *Non-representational theory & health: The health in life in space-time revealing*. Milton Park: Routledge.

Andrews, G. J. (2019). Health geographies II: The posthuman turn. *Progress in Human Geography, 43*(6), 1109–1119.

Andrews, G. J., & Duff, C. (2019a). Matter beginning to matter: On posthumanist understandings of the vital emergence of health. *Social Science & Medicine, 226*, 123–134.

Andrews, G. J., & Duff, C. (2019b). Understanding the vital emergence and expression of aging: How matter comes to matter in gerontology's posthumanist turn. *Journal of Aging Studies, 49*, 46–55.

Andrews, G. J., & Evans, J. (2008). Understanding the reproduction of health care: Towards geographies in health care work. *Progress in Human Geography, 32*(6), 759–780.

Andrews, G. J., Evans, J., & McAlister, S. (2013). 'Creating the right therapy vibe': Relational performances in holistic medicine. *Social Science & Medicine, 83*, 99–109.

Andrews, G. J., Holmes, D., Poland, B., Lehoux, P., Miller, K. L., Pringle, D., et al. (2005a). 'Airplanes are flying nursing homes': Geographies in the concepts and locales of gerontological nursing practice. *International Journal of Older People Nursing, 14*, 109–120.

Andrews, G. J., McCormack, B., & Reed, J. (2005b). The importance of place in older people's care. *International Journal of Older People Nursing, 14*, 98–99.

Andrews, G. J., & Moon, G. (2005a). Space, place, and the evidence base: Part I—An introduction to health geography. *Worldviews on Evidence-Based Nursing, 2*(2), 55–62.

Andrews, G. J., & Moon, G. (2005b). Space, place, and the evidence base: Part II—Rereading nursing environment through geographical research. *Worldviews on Evidence-Based Nursing, 2*(3), 142–156.

Andrews, G. J., & Peter, E. (2006). Moral geographies of restraint in nursing homes. *Worldviews on Evidence-Based Nursing, 3*(1), 2–7.

Andrews, G. J., & Phillips, D. R. (2002). Changing local geographies of private residential care for older people 1983–1999: Lessons for social policy in England and Wales. *Social Science & Medicine, 55*(1), 63–78.

Andrews, G. J., & Shaw, D. (2008). Clinical geography: Nursing practice and the (re) making of institutional space. *Journal of Nursing Management, 16*(4), 463–473.

Andrews, G. J., & Shaw, D. (2010). "So we started talking about a beach in Barbados": Visualization practices and needle phobia. *Social Science & Medicine, 71*, 1804–1810.

Andrews, G. J., & Shaw, D. (2012). Place visualization: Conventional or unconventional practice? *Complementary Therapies in Clinical Practice, 18*(1), 43–48.

Angus, J., Hodnett, E., & O'Brien-Pallas, L. (2003). Implementing evidence-based nursing practice: A tale of two intrapartum nursing units. *Nursing Inquiry, 10*(4), 218–228.

Anyinam, C. (1990). Alternative medicine in western industrialized countries: An agenda for medical geography. *The Canadian Geographer, 34*(1), 69–76.

Aranda, K., & Law, K. (2007). Tales of sociology and the nursing curriculum: Revisiting the debates. *Nurse Education Today, 27*(6), 561–567.

Arseneault, A. (2008). University nursing education for francophones in New Brunswick in the 1960s: The role of nuns, priests, politicians, and nurses. In J. Elliott, M. E. Stuart, & C. Toman (Eds.), *Place and practice in Canadian nursing history*. Vancouver: UBC Press.

Atherton, I., & Kyle, R. (2014). Habitat and health: Why place matters: In their second article on the social sciences, nursing lecturers Richard Kyle and Iain Atherton consider the importance of place in people's lives. *Nursing Standard, 29*(3), 24–25.

Atherton, I., & Kyle, R. (2015). Stepping outside your comfort zone. *Nursing Standard, 29*(21), 24–25.

Bach, S. (2015). Nurses across borders: The international migration of health professionals. In B. Parry, B. Greenhough, T. Brown, & I. Dyck (Eds.), *Bodies across borders: The global circulation of body parts, medical tourists and professionals*. Farnham: Ashgate.

Baer, L. D. (2003). A proposed framework for analyzing the potential replacement of international medical graduates. *Health & Place, 9*(4), 291–304.

Bandman, E. L., & Bandman, B. (2002). *Nursing ethics through the life span*. Upper Saddle River: Prentice Hall.

Barnard, A., & Sandelowski, M. (2001). Technology and humane nursing care: Irreconcilable or invented difference? *Journal of Advanced Nursing, 34*(3), 367–375.

Barnes, L., & Rudge, T. (2005). Virtual reality or real virtuality: The space of flows and nursing practice. *Nursing Inquiry, 12*(4), 306–315.

Barnett, J. R. (1988). Foreign medical graduates in New Zealand 1973–1979: A test of the exacerbation hypothesis. *Social Science & Medicine, 26*, 1038–1040.

Barnett, J. R. (1991). Geographical implications of restricting foreign medical immigration: A New Zealand case study 1976–1987. *Social Science & Medicine, 33*, 459–470.

Batnitzky, A., & McDowell, L. (2011). Migration, nursing, institutional discrimination and emotional/affective labour: Ethnicity and labour stratification in the UK National Health Service. *Social & Cultural Geography, 12*(2), 181–201.

Bauder, H. (2003). "Brain abuse", or the devaluation of immigrant labour in Canada. *Antipode, 35*(4), 699–717.

Beillon, L. M., Suserud, B. O., Karlberg, I., & Herlitz, J. (2009). Does ambulance use differ between geographic areas? A survey of ambulance use in sparsely and densely populated areas. *The American Journal of Emergency Medicine, 27*(2), 202–211.

Bender, A., Andrews, G., & Peter, E. (2010). Displacement and tuberculosis: Recognition in nursing care. *Health & Place, 16*(6), 1069–1076.

Bender, A., Clune, L., & Guruge, S. (2007). Considering place in Community Health Nursing. *Canadian Journal of Nursing Research, 39*(3), 20–35.

Bergmann, S. (2007). Theology in its spatial turn: Space, place and built environments challenging and changing the images of God. *Religion Compass, 1*(3), 353–379.

Bickes, J. T. (2000). Community health assessment using computerized geographic mapping. *Nurse Educator, 25*(4), 172–185.

Bigbee, J. L. (1993). The uniqueness of rural nursing. *The Nursing Clinics of North America, 28*(1), 131–144.

Bissell, D. (2008). Comfortable bodies: sedentary affects. Environment and Planning A, 40(7), 1697–1712.

Bissell, D. (2009). Obdurate pains, transient intensities: affect and the chronically pained body. Environment and Planning A, 41(4), 911--928.

Bissell, D. (2010). Placing affective relations: Uncertain geographies of pain. In Taking place: Non-representational theories and geography. Ashgate Publishing Ltd.

Björnsdóttir, K., Ceci, C., & Purkis, M. E. (2015). The 'right' place to care for older people: Home or institution? *Nursing Inquiry, 22*(1), 64–73.

Blake, B. J., & Bentov, L. (2001). Geographical mapping of unmarried teen births and selected sociodemographic variables. *Public Health Nursing, 18*(1), 33–39.

Bloch, J. R. (2011). Using geographical information systems to explore disparities in preterm birth rates among foreign-born and US-born black mothers. *Journal of Obstetric, Gynecologic, & Neonatal Nursing, 40*(5), 544–554.

Blumen, O. (2012). Home–work relations and the spatialization of care: Wives on the margins of the Israeli high-tech industry. *Gender, Place & Culture, 19*(1), 102–117.

Brewer, M. L., Flavell, H. L., & Jordon, J. (2017). Interprofessional team-based placements: The importance of space, place, and facilitation. *Journal of Interprofessional Care, 31*(4), 429–437.

Brodie, D. A., Andrews, G. J., Andrews, J. P., Thomas, B. G., Wong, J., & Rixon, L. (2005). Working in London hospitals: Perceptions of place in nursing students' employment considerations. *Social Science & Medicine, 61*(9), 1867–1881.

Brown, R. P., & Connell, J. (2004). The migration of doctors and nurses from South Pacific Island Nations. *Social Science & Medicine, 58*(11), 2193–2210.

Browne, A. J. (2001). The influence of liberal political ideology on nursing science. *Nursing Inquiry, 8*(2), 118–129.

Brush, B. L., & Sochalski, J. (2007). International nurse migration lessons from the Philippines. *Policy, Politics, & Nursing Practice, 8*(1), 37–46.

Buchan, J. (2001). Nurse migration and international recruitment. *Nursing Inquiry, 8*(4), 203–204.

Buchan, J., Seccombe, I., & Thomas, S. (1997). Overseas mobility of UK based nurses. *International Journal of Nursing Studies, 34*(1), 54–62.

Buchan, J., & Sochalski, J. (2004). The migration of nurses: Trends and policies. *Bulletin of the World Health Organization, 82*(8), 587–594.

Bucknall, T. (2003). The clinical landscape of critical care: Nurses' decision-making. *Journal of Advanced Nursing, 43*(3), 310–319.

Burke, T. A., McKee, J. R., Wilson, H. C., Donahue, R. M., Batenhorst, A. S., & Pathak, D. S. (2000). A comparison of time-and-motion and self-reporting methods of work measurement. *Journal of Nursing Administration, 30*(3), 118–125.

Burnett, K. (2008). The healing work of Aboriginal women in indigenous and newcomer communities. In J. Elliott, M. E. Stuart, & C. Toman (Eds.), *Place and practice in Canadian nursing history*. Vancouver: UBC Press.

Bushy, A. (2002). International perspectives on rural nursing: Australia, Canada, USA. *Australian Journal of Rural Health, 10*(2), 104–111.

Bushy, A. (2008). Conducting culturally competent rural nursing research. *Annual Review of Nursing Research, 26*(1), 221–236.

Butts, J. B., & Rich, K. L. (2019). *Nursing ethics*. Burlington: Jones & Bartlett Learning.

Cadman, L. (2009). Nonrepresentational theory/nonrepresentational geographies. In R. Kitchin & N. Thrift (Eds.), *International encyclopedia of human geography*. London: Elsevier.

Cahuas, M. C. (2019). Burned, broke, and brilliant: Latinx community workers' experiences across the Greater Toronto Area's non-profit sector. *Antipode, 51*(1), 66–86.

Cameron, E. (2007). Exhibit and point of sale: Negotiating commerce and culture at the Vancouver Art Gallery. *Social & Cultural Geography, 8*(4), 551–573.

Carolan, M., Andrews, G. J., & Hodnett, E. (2006). Writing place: A comparison of nursing research and health geography. *Nursing Inquiry, 13*(3), 203–219.

Carr-Hill, R., Shelton, T., Smith, P., Martin, S., Peacock, S., & Hardman, G. (1994). Allocating resources to health authorities: Development of a method for small area analysis of the use of inpatient services. *British Medical Journal, 109*, 1046–1049.

Castleden, H., Crooks, V. A., Hanlon, N., & Schuurman, N. (2010). Providers' perceptions of Aboriginal palliative care in British Columbia's rural interior. *Health & Social Care in the Community, 18*(5), 483–491.

Castree, N., Coe, M., Ward, K., & Samers, M. (2004a). *Spaces of work: Global capitalism and geographies of labour*. Thousand Oaks: Sage.

Castree, N., Nash, C., Badmington, N., Braun, B., Murdoch, J., & Whatmore, S. (2004b). Mapping posthumanism: An exchange. *Environment and Planning A, 36*(8), 1341–1363.

Ceci, C. (2004). Nursing, knowledge and power: A case analysis. *Social Science & Medicine, 59*(9), 1879–1889.

Cheek, J. (2004). Older people and acute care: A matter of place. *Illness, Crisis and Loss, 12*(1), 52–62.

Cho, S. H., Lee, J. Y., Mark, B. A., & Jones, C. B. (2014). Geographic mobility of Korean new graduate nurses from their first to subsequent jobs and metropolitan-nonmetropolitan differences in their job satisfaction. *Nursing Outlook, 62*(1), 22–28.

Cmiel, C. A., Karr, D. M., Gasser, D. M., Oliphant, L. M., & Neveau, A. J. (2004). Noise control: A nursing team's approach to sleep promotion: Respecting the silence creates a healthier environment for your patients. *AJN The American Journal of Nursing, 104*(2), 40–48.

Cody, W. K. (2006). Philosophical and theoretical perspectives for advanced nursing practice. Jones & Bartlett Learning.

Connell, J. (2007). Local skills and global markets? The migration of health workers from Caribbean and Pacific Island States. *Social and Economic Studies, 56*, 67–95.

Connell, J. (Ed.). (2008). *The international migration of health workers*. Milton Park: Routledge.

Connell, J. (2010a). *Migration of health workers in the Asia Pacific region*. Sydney: Human Resources for Health Knowledge Hub, University of New South Wales.

Connell, J. (2010b). *Migration and the globalisation of health care: The health worker exodus?* Cheltenham: Edward Elgar Publishing.

Connell, J. (2014). The two cultures of health worker migration: A Pacific perspective. *Social Science & Medicine, 116*, 73–81.

Connell, J., & Walton-Roberts, M. (2016). What about the workers? The missing geographies of health care. *Progress in Human Geography, 40*(2), 158–176.

Connell, J., Zurn, P., Stilwell, B., Awases, M., & Braichet, J. M. (2007). Sub-Saharan Africa: Beyond the health worker migration crisis? *Social Science & Medicine, 64*(9), 1876–1891.

Conradson, D. (2003). Doing organisational space: Practices of voluntary welfare in the city. *Environment and Planning A, 35*(11), 1975–1992.
Cormode, L., & Hughes, A. (1999). The economic geographer as a situated researcher of elites. *Geoforum, 30*, 299–300.
Cornell, P., Herrin-Griffith, D., Keim, C., Petschonek, S., Sanders, A. M., D'Mello, S., et al. (2010a). Transforming nursing workflow, part 1: The chaotic nature of nurse activities. *Journal of Nursing Administration, 40*(9), 366–373.
Cornell, P., Riordan, M., & Herrin-Griffith, D. (2010b). Transforming nursing workflow, part 2: The impact of technology on nurse activities. *Journal of Nursing Administration, 40*(10), 432–439.
Cornell, P., Riordan, M., Townsend-Gervis, M., & Mobley, R. (2011). Barriers to critical thinking: Workflow interruptions and task switching among nurses. *Journal of Nursing Administration, 41*(10), 407–414.
Corrêa, D. D. C., Andrade, M., & Coimbra, L. (2013). Liquid space: The nurse as an agent of care in offshore environment. *Journal of Nursing UFPE, 7*(12), 6907–6912.
Courtney, K. L. (2005). Visualizing nursing workforce distribution: Policy evaluation using geographic information systems. *International Journal of Medical Informatics, 74*(11), 980–988.
Coyte, P. C., & Holmes, D. (2006). Beyond the art of governmentality: Unmasking the distributional consequences of health policies. *Nursing Inquiry, 13*(2), 154–160.
Coyte, P. C., & Holmes, D. (2007). Health care technology adoption and diffusion in a social context. *Policy, Politics, & Nursing Practice, 8*(1), 47–54.
Crang, P. (1994). 'It's showtime': On the workplace geographies of display in a restaurant in southeast England. *Environment and Planning D: Society and Space, 12*(6), 675–704.
Cranston, S. (2014). Reflections on doing the expat show: Performing the global mobility industry. *Environment and Planning A, 46*(5), 1124–1138.
Crew, L., Gregson, N., & Brooks, K. (2003). The discursivities of difference: Retro retailers and the ambiguities of 'the alternative'. *Journal of Consumer Culture, 3*(1), 61–82.
Crooks, V. A., & Agarwal, G. (2008). What are the roles involved in establishing and maintaining informational continuity of care within family practice? A systematic review. *BMC Family Practice, 9*(1), 65.
Crooks, V. A., Williams, A., Stajduhar, K. I., Allan, D. E., & Cohen, R. S. (2007). The information transfer and knowledge acquisition geographies of family caregivers: An analysis of Canada's Compassionate Care Benefit. *Canadian Journal of Nursing Research, 39*(3), 36–54.
Croxson, B., Propper, C., & Perkins, A. (2001). Do doctors respond to financial incentives? UK family doctors and the GP fundholder scheme. *Journal of Public Economics, 79*(2), 375–398.
Crush, J., & Chikanda, A. (2015). South–South medical tourism and the quest for health in Southern Africa. *Social Science & Medicine, 124*, 313–320.
Cummins, S., Curtis, S., Diez-Roux, A. V., & Macintyre, S. (2007). Understanding and representing 'place' in health research: A relational approach. *Social Science & Medicine, 65*(9), 1825–1838.
Cutchin, M. (1997). Physician retention in rural communities: The perspectives of experiential place integration. *Health & Place, 3*(1), 25–41.
Cutchin, M. (2005). Spaces for inquiry into the role of place for older people's care. *Journal of Clinical Nursing, 14*, 121–129.
Daigle, J. (2008). The call of the north: Settlement nurses in the remote areas of Quebec, 1932-72. In J. Elliott, M. E. Stuart, & C. Toman (Eds.), *Place and practice in Canadian nursing history*. Vancouver: UBC Press.
Davis, D., & Walker, K. (2010). The corporeal, the social and space/place: Exploring intersections from a midwifery perspective in New Zealand. *Gender, Place and Culture, 17*(3), 377–391.
De Arellano, A. B. R. (2007). Patients without borders: The emergence of medical tourism. *International Journal of Health Services, 37*(1), 193–198.
DeGuzman, P. B., & Kulbok, P. A. (2012). Changing health outcomes of vulnerable populations through nursing's influence on neighborhood built environment: A framework for nursing research. *Journal of Nursing Scholarship, 44*(4), 341–348.

Deleuze, G., & Guattari, F. (1988). *A thousand plateaus: Capitalism and schizophrenia*. London: Bloomsbury Publishing.
DeSantis, L. (1994). Making anthropology clinically relevant to nursing care. *Journal of Advanced Nursing, 20*(4), 707–715.
Dingwall, R., Rafferty, A. M., & Webster, C. (2002). *An introduction to the social history of nursing*. Milton Park: Routledge.
Doel, M. A., & Segrott, J. (2004). Materializing complementary and alternative medicine: Aromatherapy, chiropractic, and Chinese herbal medicine in the UK. *Geoforum, 35*(6), 727–738.
Doona, M. E., Haggerty, L. A., & Chase, S. K. (1997). Nursing presence: An existential exploration of the concept. *Research and Theory for Nursing Practice, 11*(1), 3.
Ducey, A. (2007). More than a job: Meaning, affect, and training health care workers. In P. Clough & J. Halley (Eds.), *The affective turn: Theorizing the social* (pp. 187–208). Durham: Duke University Press.
Ducey, A. (2010). Technologies of caring labor: From objects to affect. In R. Parrenas & E. Boris (Eds.), *Intimate labors: Cultures, technologies, and the politics of care* (pp. 18–32). Stanford: Stanford University Press.
Duke, M., & Street, A. E. (2003). Hospital in the home: Constructions of the nursing role: A literature review. *Journal of Clinical Nursing, 12*, 852–859.
Dummer, T. J. (2008). Health geography: Supporting public health policy and planning. *CMAJ, 178*(9), 1177–1180.
Dyck, I. (1989). Integrating home and wage workplace: Women's daily lives in a Canadian suburb. *The Canadian Geographer, 33*(4), 329–341.
Dyck, I. (1990). Space, time, and renegotiating motherhood: An exploration of the domestic workplace. *Environment and Planning D: Society and Space, 8*(4), 459–483.
Dyck, I. (1999). Body troubles: Women, the workplace and negotiations of a disabled identity. In R. Butler & H. Parr (Eds.), *Mind and body spaces: Geographies of disability, illness and impairment*. Milton Park: Routledge.
Edwards, S. D. (1997). What is philosophy of nursing? *Journal of Advanced Nursing, 25*(5), 1089–1093.
Einboden, R., Rudge, T., & Varcoe, C. (2013). Image, measure, figure: A critical discourse analysis of nursing practices that develop children. *Nursing Philosophy, 14*(3), 212–222.
Elliott, J. (2008). (Re)constructing the identity of a red cross outpost nurse: The letters of Louise de Kiriline, 1927-36. In J. Elliott, M. E. Stuart, & C. Toman (Eds.), *Place and practice in Canadian nursing history*. Vancouver: UBC Press.
Elliott, J., Stuart, M. E., & Toman, C. (2008). *Place and practice in Canadian nursing history*. Vancouver: UBC Press.
Endacott, R., Boulos, M. N. K., Manning, B. R., & Maramba, I. (2009). Geographic information systems for healthcare organizations: A primer for nursing professions. *Computers Informatics Nursing, 27*(1), 50–56.
England, K. (2010). Home, work and the shifting geographies of care. *Ethics, Place and Environment, 13*(2), 131–150.
England, K. (2015). Nurses across borders: Global migration of registered nurses to the US. *Gender, Place & Culture, 22*(1), 143–156.
England, K., & Henry, C. (2013). Care work, migration and citizenship: International nurses in the UK. *Social & Cultural Geography, 14*(5), 558–574.
Ennis, C. A., & Walton-Roberts, M. (2018). Labour market regulation as global social policy: The case of nursing labour markets in Oman. *Global Social Policy, 18*(2), 169–188.
Etowa, J., Wiens, J., Bernard, W. T., & Clow, B. (2007). Determinants of Black women's health in rural and remote communities. *Canadian Journal of Nursing Research, 39*(3), 56–76.
Evans, J., & Wilton, R. (2019). Well enough to work? Social enterprise employment and the geographies of mental health recovery. *Annals of the American Association of Geographers, 109*(1), 87–103.

Everts, J. (2010). Consuming and living the corner shop: Belonging, remembering, socialising. *Social & Cultural Geography, 11*(8), 847–863.
Eyles, J., & Williams, A. (Eds.). (2008). *Sense of place, health and quality of life.* Farnham: Ashgate.
Falk-Rafael, A. (2006). Globalization and global health: Toward nursing praxis in the global community. *Advances in Nursing Science, 29*(1), 2–14.
Farmer, J., Lauder, W., Richards, H., & Sharkey, S. (2003). Dr John has gone: Assessing health professionals' contribution to remote rural community sustainability in the UK. *Social Science & Medicine, 57*(4), 673–686.
Faruque, F. S., Lofton, S. P., Doddato, T. M., & Mangum, C. (2003). Utilizing geographic information systems in community assessment and nursing research. *Journal of Community Health Nursing, 20*(3), 179–191.
Fullilove, M. T. (1996). Psychiatric implications of displacement: Contributions from the psychology of place. *American Journal of Psychiatry, 153*, 12.
Gastaldo, D., Andrews, G. J., & Khanlou, N. (2004). Therapeutic landscapes of the mind: Theorizing some intersections between health geography, health promotion and immigration studies. *Critical Public Health, 14*(2), 157–176.
Gastaldo, D., & Holmes, D. (1999). Foucault and nursing: A history of the present. *Nursing Inquiry, 6*(4), 231–240.
Gastaldo, D., & Magalhaes, L. (2010). International migration versus national health-care. *Nursing Inquiry, 17*(3), 185–185.
George, S. (2015). 'Real nursing work' versus 'charting and sweet talking': The challenges of incorporating into US urban health care settings for Indian immigrant nurses. In B. Parry, B. Greenhough, T. Brown, & I. Dyck (Eds.), *Bodies across borders: The global circulation of body parts, medical tourists and professionals*. Farnham: Ashgate.
Gesler, W. M., Hayes, M., Arcury, T. A., Skelly, A. H., Nash, S., & Soward, A. C. (2004). Use of mapping technology in health intervention research. *Nursing Outlook, 52*(3), 142–146.
Gesler, W. M., & Kearns, R. A. (2002). *Culture/place/health.* Milton Park: Routledge.
Gilmour, J. A. (2006). Hybrid space: Constituting the hospital as a home space for patients. *Nursing Inquiry, 13*(1), 16–22.
Good, C. M. (2004). *The steamer parish: The rise and fall of missionary medicine on an African frontier*. Chicago: University of Chicago Press.
Gorman-Murray, A., & Bissell, D. (2018). Mobile work, multilocal dwelling and spaces of wellbeing. *Health & Place, 51*, 232–238.
Graves, B. A. (2012a). A model for assessment of potential geographical accessibility: A case for GIS. *Online Journal of Rural Nursing and Health Care, 9*(1), 46–55.
Graves, B. A. (2012b). Geographical information systems and the analysis of health for rural populations. *Online Journal of Rural Nursing and Health Care, 8*(1), 11–12.
Greenhough, B. (2006). Tales of an island-laboratory: Defining the field in geography and science studies. *Transactions of the Institute of British Geographers, 31*(2), 224–237.
Gregson, N., Crang, M., Botticello, J., Calestani, M., & Krzywoszynska, A. (2016). Doing the 'dirty work' of the green economy: Resource recovery and migrant labour in the EU. *European Urban and Regional Studies, 23*(4), 541–555.
Gregson, N., Crewe, L., & Brooks, K. (2002). Discourse, displacement, and retail practice: Some pointers from the charity retail project. *Environment and Planning A, 34*(9), 1661–1683.
Groutsis, D. (2006). Geography and credentialism: The assessment and accreditation of overseas-trained doctors. *Health Sociology Review, 15*(1), 59–70.
Guagliardo, M. F., Ronzio, C. R., Cheung, I., Chacko, E., & Joseph, J. G. (2004). Physician accessibility: An urban case study of pediatric providers. *Health & Place, 10*(3), 273–283.
Guyatt, V. (2005). Gender performances in a service orientated workplace in Aotearoa/New Zealand. *New Zealand Geographer, 61*, 203–212.
Halford, S., & Leonard, P. (2003). Space and place in the construction and performance of gendered nursing identities. *Journal of Advanced Nursing, 42*(2), 201–208.

Halford, S., & Leonard, P. (2006a). Place, space and time: Contextualizing workplace subjectivities. *Organization Studies, 27*(5), 657–676.
Halford, S., & Leonard, P. (2006b). *Negotiating gendered identities at work: Place, space and time*. London: Palgrave.
Hall, E. (1999). Workspaces: Refiguring the disability-employment debate. In R. Butler & H. Parr (Eds.), *Mind and body spaces*. Milton Park: Routledge.
Hall, J. M. (1996). Geography of childhood sexual abuse: Women's narratives of their childhood environments. *Advances in Nursing Science, 18*(4), 29–47.
Hammond, A., Foureur, M., Homer, C. S., & Davis, D. (2013). Space, place and the midwife: Exploring the relationship between the birth environment, neurobiology and midwifery practice. *Women and Birth, 26*(4), 277–281.
Hanlon, N. T. (2001). Sense of place, organizational context and the strategic management of publicly funded hospitals. *Health Policy, 58*(2), 151–173.
Harris, A., McGillis Hall, L., Peterson, J., Price, S., Lalonde, M., Andrews, G. J., & McDonald Rencz, S. (2013). LPN perspectives on factors that affect nurse mobility in Canada. *Canadian Journal of Nursing Leadership, 26*, 70–78.
Harrison, M. E. (1995). A doctor's place: Female physicians in Mexico DF. *Health & Place, 1*(2), 101–111.
Hastings, T. (2016). Moral matters: De-romanticising worker agency and charting future directions for labour geography. *Geography Compass, 10*(7), 307–318.
Hastings, T., & MacKinnon, D. (2017). Re-embedding agency at the workplace scale: Workers and labour control in Glasgow call centres. *Environment and Planning A: Economy and Space, 49*(1), 104–120.
Hendrich, A., Chow, M. P., Bafna, S., Choudhary, R., Heo, Y., & Skierczynski, B. A. (2009). Unit-related factors that affect nursing time with patients: Spatial analysis of the time and motion study. *HERD, 2*(2), 5.
Hendrich, A., Chow, M. P., Skierczynski, B. A., & Lu, Z. (2008). A 36-hospital time and motion study: How do medical-surgical nurses spend their time? *The Permanente Journal, 12*(3), 25.
Herdman, E. A. (2004). Globalization, internationalization and nursing. *Nursing & Health Sciences, 6*(4), 237–238.
Herron, R. V., & Skinner, M. W. (2012). Farmwomen's emotional geographies of care: A view from rural Ontario. *Gender, Place & Culture, 19*(2), 232–248.
Hodgins, H. J., & Wuest, J. (2007). Uncovering factors affecting use of the emergency department for less urgent health problems in urban and rural areas. *Canadian Journal of Nursing Research, 39*(3), 78–102.
Hodnett, E. D., Downe, S., Edwards, N., & Walsh, D. (2005). Home-like versus conventional institutional settings for birth. *Birth, 32*(2), 151–151.
Hodnett, E. D., Stremler, R., Weston, J. A., & McKeever, P. (2009). Re-conceptualizing the hospital labor room: The PLACE (pregnant and laboring in an ambient clinical environment) pilot trial. *Birth, 36*(2), 159–166.
Hoffart, M. B., & Keene, E. P. (1998). The benefits of visualization: Research suggests that visualization promotes relaxation, enhances sleep, reduces pain, and increases creativity. *The American Journal of Nursing, 98*(12), 44–47.
Holden, P., & Littlewood, J. (Eds.). (2015). *Anthropology and nursing*. Milton Park: Routledge.
Holland, K. (1999). A journey to becoming: The student nurse in transition. *Journal of Advanced Nursing, 29*(1), 229–236.
Holmes, D. (2001). From iron gaze to nursing care: Mental health nursing in the era of panopticism. *Journal of Psychiatric and Mental Health Nursing, 8*(1), 7–15.
Holmes, D., Gastaldo, D., & Perron, A. (2007a). Paranoid investments in nursing: A schizoanalysis of the evidence-based discourse. *Nursing Philosophy, 8*(2), 85–91.
Holmes, D., O'byrne, P., & Coyte, P. C. (2007b). M3S squad: Considering cyborgs for future sexual health prevention programs. *Body, Space & Technology, 7*(1), 1–9.
Holmes, D., O'Byrne, P., & Murray, S. J. (2010). Faceless sex: Glory holes and sexual assemblages. *Nursing Philosophy, 11*(4), 250–259.

Horowitz, M. D. (2007). Medical tourism-health care in the global economy. *Physician Executive, 33*(6), 24.

Horowitz, M. D., Rosensweig, J. A., & Jones, C. A. (2007). Medical tourism: Globalization of the healthcare marketplace. *Medscape General Medicine, 9*(4), 33.

Hudson, C. G. (2012). Disparities in the geography of mental health: Implications for social work. *Social Work, 57*(2), 107–119.

Hudson, R. (2001). *Producing places*. New York: Guilford Press.

Hujala, A., & Rissanen, S. (2011). Organization aesthetics in nursing homes. *Journal of Nursing Management, 19*(4), 439–448.

Hurst, K. (2008). UK ward design: Patient dependency, nursing workload, staffing and quality—An observational study. *International Journal of Nursing Studies, 45*(3), 370–381.

Huynh, T., & Alderson, M. (2009). Concept analysis of human ecology. *Nursing Forum, 44*(2), 115–128.

Iredale, R., Jones, L., Gray, J., & Deaville, J. (2005). 'The edge effect': An explanatory study of some factors affecting referrals to cancer general services in rural Wales. *Health & Place, 11*(3), 197–204.

Jacob, J. M. (2017). *Doctors and rules: A sociology of professional values*. Milton Park: Routledge.

James, A. (2011). Work-life (im)'balance' and its consequences for everyday learning and innovation in the New Economy: Evidence from the Irish IT sector. *Gender, Place & Culture, 18*(5), 655–684.

James, A. (2017). *Work-life advantage: Sustaining regional learning and innovation*. Hoboken: John Wiley & Sons.

Jennings, M. (2008). 'Healing of bodies, salvation of souls': Missionary medicine in colonial Tanganyika, 1870s-1939. *Journal of Religion in Africa, 38*(1), 27–56.

Jensen, A., & Curtis, M. (2008). A descriptive qualitative study of student learning in a psychosocial nursing class infused with art, literature, music, and film. *International Journal of Nursing Education Scholarship, 5*(1), 1–9.

Johns, C., & Freshwater, D. (2009). *Transforming nursing through reflective practice*. Hoboken: John Wiley & Sons.

Johnston, R., Crooks, V. A., Snyder, J., & Kingsbury, P. (2010). What is known about the effects of medical tourism in destination and departure countries? A scoping review. *International Journal for Equity in Health, 9*(1), 24.

Jones, A. C. (2004). Transference and countertransference. *Perspectives in Psychiatric Care, 40*(1), 13–19.

Jones, C. B., & Sherwood, G. (2014). The globalization of the nursing workforce: Pulling the pieces together. *Nursing Outlook, 62*(1), 59–63.

Jones, M. (2009). Phase space: Geography, relational thinking, and beyond. *Progress in Human Geography, 33*(4), 487–506.

Joseph, A. E., & Phillips, D. R. (1984). *Accessibility and utilization: Geographical perspectives on health care delivery*. Thousand Oaks: Sage.

Judd, D. M., & Sitzman, K. (2014). *History of American nursing: Trends and eras*. Sudbury: Jones & Bartlett Publishers.

Kakar, S. (1996). Leprosy in British India, 1860–1940: Colonial politics and missionary medicine. *Medical History, 40*(2), 215–230.

Kalusa, W. T. (2007). Language, medical auxiliaries, and the re-interpretation of missionary medicine in Colonial Mwinilunga, Zambia, 1922–51. *Journal of Eastern African Studies, 1*(1), 57–78.

Kazanjian, A., & Pagliccia, N. (1996). Key factors in physicians' choice of practice location: Findings from a survey of practitioners and their spouses. *Health & Place, 2*(1), 27–34.

Kearns, R., & Moon, G. (2002). From medical to health geography: Novelty, place and theory after a decade of change. *Progress in Human Geography, 26*(5), 605–625.

Kearns, R. A. (1993). Place and health: Towards a reformed medical geography. *The Professional Geographer, 45*(2), 139–147.

Kearns, R. A., & Barnett, J. R. (1997). Consumerist ideology and the symbolic landscapes of private medicine. *Health & place , 3*(3), 171–180.

Kearns, R. A., & Barnett, J. R. (2000). "Happy Meals" in the Starship Enterprise: Interpreting a moral geography of health care consumption. *Health & Place, 6*(2), 81–93.
Kearns, R. A., Barnett, J. R., & Newman, D. (2003). Placing private health care: Reading Ascot hospital in the landscape of contemporary Auckland. *Social Science & Medicine, 56*(11), 2303–2315.
Keller, L. O., Strohschein, S., Lia-Hoagberg, B., & Schaffer, M. A. (2004). Population-based public health interventions: Practice-based and evidence-supported. Part I. *Public Health Nursing, 21*(5), 453–468.
King, E. C., Holliday, P. J., & Andrews, G. J. (2018a). Care challenges in the bathroom: The views of professional care providers working in clients' homes. *Journal of Applied Gerontology, 37*(4), 493–515.
King, N., Horrocks, C., & Brooks, J. (2018b). *Interviews in qualitative research*. Thousand Oaks: Sage.
Kingma, M. (2001). Nursing migration: Global treasure hunt or disaster-in-the-making? *Nursing Inquiry, 8*(4), 205–212.
Kingma, M. (2003). Economic incentive in community nursing: Attraction, rejection or indifference. *Human Resources for Health, 1*(2), 1–12.
Kingma, M. (2006). *Nurses on the move: Migration and the global health care economy*. Ithica: Cornell University Press.
Kingma, M. (2007). Nurses on the move: A global overview. *Health Services Research, 42*(3), 1281–1298.
Kingma, M. (2008). Nurses on the move: Diversity and the work environment. *Contemporary Nurse, 28*(1–2), 198–206.
Kitto, S., Nordquist, J., Peller, J., Grant, R., & Reeves, S. (2013). The disconnections between space, place and learning in interprofessional education: An overview of key issues. *Journal of Interprofessional Care, 27*(S2), 5–8.
Kline, D. S. (2003). Push and pull factors in international nurse migration. *Journal of Nursing Scholarship, 35*(2), 107–111.
Kobayashi, A. (1994). *Women, work, and place*. Montreal: McGill-Queen's Press-MQUP.
Kovner, C. T., Corcoran, S. P., & Brewer, C. S. (2011). The relative geographic immobility of new registered nurses calls for new strategies to augment that workforce. *Health Affairs, 30*(12), 2293–2300.
Kyle, R., & Atherton, I. (2014). Habitat and health: Why place matters. *Nursing Standard, 29*(3), 24–25.
Kyle, R. G., & Atherton, I. M. (2016). Biogeography as critical nursing pedagogy: Breathing life into nurse education. *Nurse Education in Practice, 20*, 76–79.
Kyle, R. G., Atherton, I. M., Kesby, M., Sothern, M., & Andrews, G. (2016). Transfusing our lifeblood: Reframing research impact through inter-disciplinary collaboration between health geography and nurse education. *Social Science & Medicine, 168*, 257–264.
Laditka, J. N. (2004). Physician supply, physician diversity and outcomes of primary health care for older persons in the United States. *Health & Place, 10*(3), 231–244.
Lapum, J., Chen, S., Peterson, J., Leung, D., & Andrews, G. J. (2009). The place of nursing in primary health care. In V. Crooks & G. J. Andrews (Eds.), *Primary health care: People, practice, place*. Farnham: Ashgate.
Lapum, J., Fredericks, S., Beanlands, H., McCay, E., Schwind, J., & Romaniuk, D. (2012a). A cyborg ontology in health care: Traversing into the liminal space between technology and person-centred practice. *Nursing Philosophy, 13*(4), 276–288.
Lee, S. J., & Park, J. M. (2006). [Evaluation of visiting nursing care using geographical information system (GIS) technology]. *Journal of Korean Academy of Nursing, 36*(6), 1042–1054.
Lehoux, P. (2008). Why examining the desirability of health technology matters. *Healthcare Policy, 3*(3), 29.
Lehoux, P., Daudelin, G., Poland, B., Andrews, G. J., & Holmes, D. (2007). Designing a better place for patients: Professional struggles surrounding satellite and mobile dialysis units. *Social Science & Medicine, 65*(7), 1536–1548.

Lehoux, P., Poland, B., Daudelin, G., Holmes, D., & Andrews, G. (2008). Displacement and emplacement of health technology: Making satellite and mobile dialysis units closer to patients? *Science, Technology, & Human Values, 33*(3), 364–392.
Leipert, B., & Anderson, E. (2012). Rural nursing education: A photovoice perspective. *Rural and Remote Health, 12*, 2061.
Leipert, B., & Reutter, L. (1998). Women's health and community health nursing practice in geographically isolated settings: A Canadian perspective. *Health Care for Women International, 19*(6), 575–588.
Levin, K. A., Davies, C. A., Douglas, G. V., & Pitts, N. B. (2010). Urban-rural differences in dental caries of 5-year old children in Scotland. *Social Science & Medicine, 71*(11), 2020–2027.
Liaschenko, J. (1994). The moral geography of home care. *Advances in Nursing Science, 17*, 16–25.
Liaschenko, J. (1996a). A sense of place for patients: Living and dying. *Home Care Provider, 1*, 270–272.
Liaschenko, J. (1996b). Home is different: On place and ethics. *Home Care Provider, 1*(1), 49–50.
Liaschenko, J. (1997). Ethics and the geography of the nurse–patient relationship: Spatial vulnerabilities and gendered space. *Scholarly Inquiry for Nursing Practice, 11*, 45–59.
Liaschenko, J. (2000). The moral geography of care work. *Home Care Provider, 5*(4), 126–127, 132.
Liaschenko, J. (2001). Nursing work, housekeeping issues, and the moral geography of home care. In D. N. Weisstub, D. C. Thomasma, S. Gauthier, & G. F. Tomossy (Eds.), *Aging: Caring for our elders* (pp. 123–137). New York: Springer.
Liaschenko, J. (2003). At home with illness: Moral understandings and moral geographies. *Ethica, 15*(2), 71–82.
Liaschenko, J., Peden-McAlpine, C., & Andrews, G. J. (2011). Institutional geographies in dying: Nurses' actions and observations on dying spaces inside and outside intensive care units. *Health & Place, 17*(3), 814–821.
Liaschenko, J., & Peter, E. (2004). Nursing ethics and conceptualizations of nursing: Profession, practice and work. *Journal of Advanced Nursing, 46*(5), 488–495.
Lin, G., Burns, P. A., & Nochajski, T. H. (1997). The geographic distribution of nurse practitioners in the United States. *Applied Geographic Studies, 1*(4), 287–301.
Liu, W., Manias, E., & Gerdtz, M. (2014). The effects of physical environments in medical wards on medication communication processes affecting patient safety. *Health & Place, 26*, 188–198.
Lock, L. R., & Gibb, H. J. (2003). The power of place. *Midwifery, 19*, 132–139.
Lovell, S. A., Kearns, R. A., & Rosenberg, M. W. (2011). Community capacity building in practice: Constructing its meaning and relevance to health promoters. *Health & Social Care in the Community, 19*(5), 531–540.
Lunt, N., & Carrera, P. (2010). Medical tourism: Assessing the evidence on treatment abroad. *Maturitas, 66*(1), 27–32.
Malone, R. (2003). Distal nursing. *Social Science & Medicine, 56*, 2317–2326.
Mapedzahama, V., Rudge, T., West, S., & Perron, A. (2012). Black nurse in white space? Rethinking the in/visibility of race within the Australian nursing workplace. *Nursing Inquiry, 19*(2), 153–164.
Marchant, C., & O'Donohoe, S. (2019). Intercorporeality and the emerging adult-smartphone assemblage. *Information Technology & People, 32*(2), 453–474.
Marshall, E. S. (2008). Home as place for healing. *Advances in Nursing Science, 31*(3), 259–267.
Mason, D. J., Leavitt, J. K., & Chaffee, M. W. (2013). *Policy and politics in nursing and healthcare-revised reprint*. New York: Elsevier Health Sciences.
Mayer, J. D. (1979). Seattle's paramedic program: Geographical distribution, response times, and mortality. *Social Science & Medicine. Part D: Medical Geography, 13*(1), 45–51.
McCaffrey, R., & Locsin, R. C. (2002). Music listening as a nursing intervention: A symphony of practice. *Holistic Nursing Practice, 16*(3), 70–77.

McDowell, L., Batnitzky, A., & Dyer, S. (2007). Division, segmentation, and interpellation: The embodied labors of migrant workers in a greater London hotel. *Economic Geography, 83*(1), 1–25.

McDowell, L., & Court, G. (1994). Performing work: Bodily representations in merchant banks. *Environment and Planning D: Society and Space, 12*, 727–750.

McGillis Hall, L., Peterson, J., Price, S., Andrews, G. J., Lalonde, M., Johnson, S., et al. (2013). I was never recruited: Challenges in cross-Canada nurse mobility. *Canadian Journal of Nursing Leadership, 26*, 29–40.

McGillis Hall, L., Pink, G. H., Jones, C., Leatt, P., Gates, M., Pink, L., et al. (2009). Gone south: Why Canadian nurses migrate to the United States. *Healthcare Policy, 4*(4), 91.

McHugh, K. E. (2009). Movement, memory, landscape: An excursion in non-representational thought. *GeoJournal, 74*, 209–218.

McIntosh, P. (2008). Poetics and space: Developing a reflective landscape through imagery and human geography. *Reflective Practice, 9*(1), 69–78.

McKay, M. (2008). Region, faith, and health: The development of Winnipeg's Visiting Nursing Agencies, 1897-1926. In J. Elliott, M. E. Stuart, & C. Toman (Eds.), *Place and practice in Canadian nursing history*. Vancouver: UBC Press.

McKeever, P., & Coyte, P. C. (2002). Here, there and everywhere. *University of Toronto Bulletin A, 16*, 25.

McKeever, P., Stevens, B., Miller, K. L., MacDonell, J. W., Gibbins, S., Guerriere, D., ... & Coyte, P. C. (2002). Home versus hospital breastfeeding support for newborns: a randomized controlled trial. Birth, 29(4), 258–265.

McMeekin, P., Gray, J., Ford, G. A., Duckett, J., & Price, C. I. (2014). A comparison of actual versus predicted emergency ambulance journey times using generic Geographic Information System software. *Emergency Medicine Journal, 31*(9), 758–762.

Mesman, J. (2009). The geography of patient safety: A topical analysis of sterility. *Social Science & Medicine, 69*(12), 1705–1712.

Mesman, J. (2012). Moving in with care about patient safety as a spatial achievement. *Space and Culture, 15*(1), 31–43.

Mickley, J. R., Carson, V., & Soeken, K. L. (1995). Religion and adult mental health: State of the science in nursing. *Issues in Mental Health Nursing, 16*(4), 345–360.

Miller, K. L., McKeever, P., & Coyte, P. C. (2003). Recruitment issues in healthcare research: The situation in home care. *Health & Social Care in the Community, 11*(2), 111–123.

Milligan, C. (2017). *Geographies of care: Space, place and the voluntary sector: Space, place and the voluntary sector*. Milton Park: Routledge.

Mohr, A. (2009). Missionary medicine and Akan therapeutics: Illness, health and healing in southern Ghana's Basel Mission, 1828-1918. *Journal of Religion in Africa, 39*(4), 429–461.

Montgomery, P. (2001). Shifting meaning of asylum. *Journal of Advanced Nursing, 33*(4), 425–431.

Moon, G. R. A. H. A. M., & Brown, T. (1998). Place, space and health service reform. Kearns R and Gesler R Putting health in its Place. Syracuse University Press, Syracuse, NY, 270–288.

Moon, G., Mohan, J., Twigg, L., McGrath, K., & Pollock, A. (2002). Catching waves: The historical geography of the general practice fundholding initiative in England and Wales. *Social Science & Medicine, 55*(12), 2201–2213.

Morris, S. (1998). *Health economics for nurses: An introductory guide*. Upper Saddle River: Prentice Hall PTR.

Morrison, I., Lewis, B., Liaw, S. T., Deveny, E. (2004). Modelling the clinical processes of prescribing (MCPOP)—Information, clinical workflow and processes. In *HIC 2004: Proceedings*, p. 194.

Moss, M., & Schell, M. (2004). GIS(c): A scientific framework and methodological tool for nursing research. *Advances in Nursing Science, 27*(2), 150–159.

Mugavin, M. E. (2014). Millennials usher us in: Nursing and the emerging virtual environment of care. *Journal of Nursing Education and Practice, 4*(9), 69.

Nairn, S., & Pinnock, D. (2017). Pierre Bourdieu: Expanding the scope of nursing research and practice. *Nursing Philosophy, 18*(4), e12167.

Nelson, S. (2001). *Say little, do much: Nursing, nuns, and hospitals in the nineteenth century*. Philadelphia: University of Pennsylvania Press.

Nelson, S., Verma, S., Hall, L. M., Gastaldo, D., & Janjua, M. (2011). The shifting landscape of immigration policy in Canada: Implications for health human resources. *Healthcare Policy, 7*(2), 60.

Neuwelt, P. M., Kearns, R. A., & Browne, A. J. (2015). The place of receptionists in access to primary care: Challenges in the space between community and consultation. *Social Science & Medicine, 133*, 287–295.

Newman, M. A. (1999). The rhythm of relating in a paradigm of wholeness. *The Journal of Nursing Scholarship, 31*(3), 227–230.

O'Kelly, G. (1998). Countertransference in the nurse-patient relationship: A review of the literature. *Journal of Advanced Nursing, 28*(2), 391–397.

O'Reilly, K. (2011). "We are not contractors": Professionalizing the interactive service work of NGOs in Rajasthan, India. *Economic Geography, 87*(2), 207–226.

Oandasan, I. F., Gotlib Conn, L., Lingard, L., Karim, A., Jakubovicz, D., Whitehead, C., et al. (2009). The impact of space and time on interprofessional teamwork in Canadian primary health care settings: Implications for health care reform. *Primary Health Care Research & Development, 10*(2), 151–162.

Ong, M. E., Ng, F. S., Overton, J., Yap, S., Andresen, D., Yong, D. K., et al. (2009). Geographic-time distribution of ambulance calls in Singapore: Utility of geographic information system in ambulance deployment (CARE 3). *Annals Academy of Medicine Singapore, 38*(3), 184.

Ovesen, J., & Trankell, I. B. (2010). *Cambodians and their doctors: A medical anthropology of colonial and post-colonial Cambodia*. Copenhagen: Nias Press.

Paley, J. (2014). Cognition and the compassion deficit: The social psychology of helping behaviour in nursing. *Nursing Philosophy, 15*(4), 274–287.

Papazisis, G., Nicolaou, P., Tsiga, E., Christoforou, T., & Sapountzi-Krepia, D. (2014). Religious and spiritual beliefs, self-esteem, anxiety, and depression among nursing students. *Nursing & Health Sciences, 16*(2), 232–238.

Pardo Mora, Y. Y., & González Ballesteros, M. M. (2007). Space and territory in community nursing practice. *Aquichán, 7*(2), 189–198.

Parr, H. (1998a). The politics of methodology in 'post-medical geography': Mental health research and the interview. *Health & Place, 4*(4), 341–353.

Parr, H. (1998b). Mental health, ethnography and the body. *Area, 30*(1), 28–37.

Patterson, E. F. (1998). The philosophy and physics of holistic health care: Spiritual healing as a workable interpretation. *Journal of Advanced Nursing, 27*(2), 287–293.

Peleg, K., & Pliskin, J. S. (2004). A geographic information system simulation model of EMS: Reducing ambulance response time. *The American Journal of Emergency Medicine, 22*(3), 164–170.

Peter, E. (2002). The history of nursing in the home: Revealing the significance of place in the expression of moral agency. *Nursing Inquiry, 9*(2), 65–72.

Peter, E. (2003). The home as a site of care: Ethical implications for professional boundaries and privacy. *Ethica, 15*(2), 57–69.

Peter, E., & Liaschenko, J. (2004). Paradoxes of proximity: A spatio-temporal analysis of moral distress and moral ambiguity. *Nursing Inquiry, 11*(4), 218–225.

Peter, E., Mohammed, S., & Simmonds, A. (2014). Narratives of aggressive care: Knowledge, time & responsibility. *Nursing Ethics, 21*(4), 461–472.

Pile, S. (2010). Emotions and affect in recent human geography. *Transactions of the Institute of British Geographers, 35*(1), 5–20.

Plouffe, L., & Kalache, A. (2010). Towards global age-friendly cities: Determining urban features that promote active aging. *Journal of Urban Health, 87*(5), 733–739.

Poland, B., Lehoux, P., Holmes, D., & Andrews, G. (2005). How place matters: Unpacking technology and power in health and social care. *Health & Social Care in the Community, 13*(2), 170–180.

Porter, S., & Ryan, S. (1996). Breaking the boundaries between nursing and sociology: A critical realist ethnography of the theory-practice gap. *Journal of Advanced Nursing, 24*(2), 413–420.

Pottie-Sherman, Y., & Lynch, N. (2019). Gaming on the edge: Mobile labour and global talent in Atlantic Canada's video game industry. *The Canadian Geographer, 63*(3), 425–439.

Power, A. (2008). Caring for independent lives: Geographies of caring for young adults with intellectual disabilities. *Social Science & Medicine, 67*(5), 834–843.

Pryce, A. (2004). 'Only odd people wore suede shoes': 1 careers and sexual identities of men attending a sexual health clinic. *Nursing Inquiry, 11*(4), 258–270.

Purkis, M. E. (1996). Nursing in quality space: technologies governing experiences of care. *Nursing Inquiry, 3*(2), 101–111.

Purkis, M. E. (2001). Managing home nursing care: Visibility, accountability and exclusion. *Nursing Inquiry, 8*(3), 141–150.

Pykett, J., & Enright, B. (2016). Geographies of brain culture: Optimism and optimisation in workplace training programmes. *Cultural Geographies, 23*(1), 51–68.

Quiney, L. (2008). "Suitable young women": Red Cross nursing pioneers and the crusade for healthy living in Manitoba, 1920-30. In J. Elliott, M. E. Stuart, & C. Toman (Eds.), *Place and practice in Canadian nursing history*. Vancouver: UBC Press.

Racine, L., & Perron, A. (2012). Unmasking the predicament of cultural voyeurism: A postcolonial analysis of international nursing placements. *Nursing Inquiry, 19*(3), 190–201.

Radcliffe, P. (1999). Geographical mobility, children and career progress in British professional nursing. *Journal of Advanced Nursing, 30*(3), 758–768.

Rafferty, A. M. (1995). Art, science and social science in nursing: Occupational origins and disciplinary identity. *Nursing Inquiry, 2*(3), 141–148.

Raingruber, B. J. (2001). Settling into and moving in a climate of care: Styles and patterns of interaction between nurse psychotherapists and clients. *Perspectives in Psychiatric Care, 37*(1), 15–27.

Rapport, F., Doel, M. A., & Elwyn, G. (2007). Snapshots and snippets: General practitioners' reflections on professional space. *Health & Place, 13*(2), 532–544.

Relph, E. (1976). *Place and placelessness*. London: Pion.

Richardson, D. B., Volkow, N. D., Kwan, M. P., Kaplan, R. M., Goodchild, M. F., & Croyle, R. T. (2013). Spatial turn in health research. *Science, 339*(6126), 1390–1392.

Riley, R., & Manias, E. (2002). Foucault could have been an operating room nurse. *Journal of Advanced Nursing, 39*(4), 316–324.

Rogaly, B., & Qureshi, K. (2017). 'That's where my perception of it all was shattered': Oral histories and moral geographies of food sector workers in an English city region. *Geoforum, 78*, 189–198.

Ross, S. J., Polsky, D., & Sochalski, J. (2005). Nursing shortages and international nurse migration. *International Nursing Review, 52*(4), 253–262.

Roush, C. V., & Cox, J. E. (2000). The meaning of home: How it shapes the practice of home and hospice care. *Home Healthcare Nurse, 18*(6), 388–394.

Rowles, G. D. (2000). Habituation and being in place. *The Occupational Therapy Journal of Research, 20*(1_Suppl), 52S–67S.

Rudge, T. (2013). Desiring productivity: Nary a wasted moment, never a missed step! *Nursing Philosophy, 14*(3), 201–211.

Rutakumwa, W., & Krogman, N. (2007). Women's health in rural Uganda: Problems, coping strategies, and recommendations for change. *Canadian Journal of Nursing Research, 39*(3), 104–125.

Rutherdale, M. (2008). Cleansers, cautious caregivers, and optimistic adventurers: A proposed typology of Arctic Canadian Nurses, 1945-70. In J. Elliott, M. E. Stuart, & C. Toman (Eds.), *Place and practice in Canadian nursing history*. Vancouver: UBC Press.

Sah, S., & Fugh-Berman, A. (2013). Physicians under the influence: Social psychology and industry marketing strategies. *The Journal of Law, Medicine & Ethics, 41*(3), 665–672.

Sandal, L. F., Thorlund, J. B., Moore, A. J., Ulrich, R. S., Dieppe, P. A., & Roos, E. M. (2019). Room for improvement: A randomised controlled trial with nested qualitative interviews on space, place and treatment delivery. *British Journal of Sports Medicine, 53*(6), 359–367.

Sandelowski, M. (1996). Tools of the trade: Analyzing technology as object in nursing. *Research and Theory for Nursing Practice, 10*(1), 5–16.

Sandelowski, M. (1999). Troubling distinctions: A semiotics of the nursing/technology relationship. *Nursing Inquiry, 6*(3), 198–207.

Sandelowski, M. (2002). Visible humans, vanishing bodies, and virtual nursing: Complications of life, presence, place, and identity. *Advances in Nursing Science, 24*(3), 58–70.

Savage, J. (1997). Gestures of resistance: The nurse's body in contested space. *Nursing Inquiry, 4*(4), 237–245.

Scull, A. (Ed.). (2015). *Madhouses, mad-doctors, and madmen: The social history of psychiatry in the Victorian era*. Philadelphia: University of Pennsylvania Press.

Secker, B., Goldenberg, M. J., Gibson, B. E., Wagner, F., Parke, B., Breslin, J., et al. (2006). Just regionalisation: Rehabilitating care for people with disabilities and chronic illnesses. *BMC Medical Ethics, 7*(1), 9.

Seto-Nielsen, L. S., Angus, J. E., Gastaldo, D., Howell, D., & Husain, A. (2013). Maintaining distance from a necessary intrusion: A postcolonial perspective on dying at home for Chinese immigrants in Toronto, Canada. *European Journal of Oncology Nursing, 17*(5), 649–656.

Shattell, M. M., Andes, M., & Thomas, S. P. (2008). How patients and nurses experience the acute care psychiatric environment. *Nursing Inquiry, 15*(3), 242–250.

Shreffler, J. M. (1996). An ecological view of the rural environment: Levels of influence on access to health care. *Advances in Nursing Science, 18*(4), 48–59.

Skelly, A. H., Arcury, T. A., Gesler, W. M., Cravey, A. J., Dougherty, M. C., Washburn, S. A., & Nash, S. (2002). Sociospatial knowledge networks: Appraising community as place. *Research in Nursing & Health, 25*(2), 159–170.

Skinner, M. W. (2014). Ageing, place and voluntarism: Towards a geographical perspective on third sector organisations and volunteers in ageing communities. *Voluntary Sector Review, 5*(2), 161–179.

Skott, C., & Lundgren, S. M. (2009). Complexity and contradiction: Home care in a multicultural area. *Nursing Inquiry, 16*(3), 223–231.

Sloan, R. P., Bagiella, E., & Powell, T. (1999). Religion, spirituality, and medicine. *The Lancet, 353*(9153), 664–667.

Smith, B. J., & Cantrell, P. J. (1988). DISTANCE: In nurse-patient encounters. *Journal of Psychosocial Nursing and Mental Health Services, 26*(2), 22–26.

Snyder, L. H. (1973). An exploratory study of patterns of social interaction, organization, and facility design in three nursing homes. *The International Journal of Aging and Human Development, 4*(4), 319–333.

Solberg, S. M., & Way, C. (2007). The importance of geography and health in nursing research. *Canadian Journal of Nursing Research, 39*(3), 13–18.

Solomon, H. (2011). Affective journeys: The emotional structuring of medical tourism in India. *Anthropology and Medicine, 18*(1), 105–118.

Stuart, M. (2008). Social sisters: A feminist analysis of the discourses of Canadian Military Nurse Helen Fowlds, 1915-18. In J. Elliott, M. E. Stuart, & C. Toman (Eds.), *Place and practice in Canadian nursing history*. Vancouver: UBC Press.

Stuart, M. E., Elliott, J., & Toman, C. (2008). *Place and practice in Canadian nursing history*. Vancouver: UBC Press.

Tarlier, D. S., Browne, A. J., & Johnson, J. (2007). The influence of geographical and social distance on nursing practice and continuity of care in a remote First Nations community. *Canadian Journal of Nursing Research, 39*(3), 126–148.

Taylor, B., Mathers, J., Atfield, T., & Parry, J. (2011). What are the challenges to the Big Society in maintaining lay involvement in health improvement, and how can they be met? *Journal of Public Health, 33*(1), 5–10.

Thomas, B. S. P. (2013a). The global phenomenon of urbanization and its effects on mental health. *Issues in Mental Health Nursing, 34*(3), 139–140.
Thomas, E. A. (2011). Cushioning, weathering, place, and community in rural grandmother-mothering. *Issues in Mental Health Nursing, 32*(5), 291–300.
Thomas, E. A. (2013b). Place knowing of persons and populations restoring the place work of nursing. *Journal of Holistic Nursing, 31*(4), 267–275.
Thompson, L. (2008). The role of nursing in governmentality, biopower and population health: Family health nursing. *Health & Place, 14*(1), 76–84.
Thompson, L., & Bidwell, S. (2015). Space, time, and emotion in the community pharmacy. *Health & Place, 34*, 251–256.
Thompson, M., & Walton-Roberts, M. (2019). International nurse migration from India and the Philippines: The challenge of meeting the sustainable development goals in training, orderly migration and healthcare worker retention. *Journal of Ethnic and Migration Studies, 45*(14), 2583–2599.
Thomson, H. (2008). A dose of realism for healthy urban policy: Lessons from area-based initiatives in the UK. *Journal of Epidemiology & Community Health, 62*(10), 932–936.
Thrift, N. (2004a). Summoning life. In M. Goodwin, P. Crang, & P. Cloke (Eds.), *Envisioning human geographies* (pp. 81–103). Milton Park: Routledge.
Thrift, N. (2004b). Intensities of feeling: Towards a spatial politics of affect. *Geografiska Annaler B, 86*, 57–78.
Thrift, N. (2008). *Non-representational theory: Space, politics, affect*. London: Routledge.
Timmermans, S., Bowker, G. C., & Star, S. L. (1998). The architecture of difference: Visibility, control and comparability in building a nursing interventions classification. In M. Berg & A. Mol (Eds.), *A differences in medicine: Unraveling practices, techniques, and bodies* (pp. 202–225). Durham: Duke University Press.
Toman, C. (2008). "A loyal body of empire citizens": Military nurses and identity at Lemnos and Salonika, 1915-17. In J. Elliott, M. E. Stuart, & C. Toman (Eds.), *Place and practice in Canadian nursing history*. Vancouver: UBC Press.
Tuan, Y. F. (1977). *Space and place: The perspective of experience*. Minneapolis: University of Minnesota Press.
Tucakovic, M. (1999). Working with the notion of soul in nursing. *Australian Journal of Holistic Nursing, 6*(2), 39.
Twigg, J. (1999). Regional variation in Russian medical insurance: Lessons from Moscow and Nizhny Novgorod. *Health & Place, 5*(3), 235–245.
Tyrawski, J., & DeAndrea, D. C. (2015). Pharmaceutical companies and their drugs on social media: A content analysis of drug information on popular social media sites. *Journal of Medical Internet Research, 17*(6), e130.
Uyeno, D. H., & Seeberg, C. (1984). A practical methodology for ambulance location. *Simulation, 43*(2), 79–87.
Van der Eijk, P. J. (2005). *Medicine and philosophy in classical antiquity: Doctors and philosophers on nature, soul, health and disease*. Cambridge: Cambridge University Press.
Van Ingen, C. (2003). Geographies of gender, sexuality and race reframing the focus on space in sport sociology. *International Review for the Sociology of Sport, 38*(2), 201–216.
Vandemark, L. M. (2007). Promoting the sense of self, place, and belonging in displaced persons: The example of homelessness. *Archives of Psychiatric Nursing, 21*(5), 241–248.
Vannini, P. (2009). nonrepresentational theory and symbolic interactionism: Shared perspectives and missed articulations. *Symbolic Interaction, 32*(3), 282–286.
Walton-Roberts, M. (2012). Contextualizing the global nursing care chain: International migration and the status of nursing in Kerala, India. *Global Networks, 12*(2), 175–194.
Walton-Roberts, M. (2015). International migration of health professionals and the marketization and privatization of health education in India: From push–pull to global political economy. *Social Science & Medicine, 124*, 374–382.

Walton-Roberts, M. (2019). Occupational (im) mobility in the global care economy: The case of foreign-trained nurses in the Canadian context. *Journal of Ethnic and Migration Studies, 46*, 1–16.

Walton-Roberts, M. (2020). Intermediaries and transnational regimes of skill: Nursing skills and competencies in the context of international migration. *Journal of Ethnic and Migration Studies*. https://doi.org/10.1080/1369183X.2020.1731988.

Walton-Roberts, M., Bhutani, S., & Kaur, A. (2017a). Care and global migration in the nursing profession: A north Indian perspective. *Australian Geographer, 48*(1), 59–77.

Walton-Roberts, M., Runnels, V., Rajan, S. I., Sood, A., Nair, S., Thomas, P., et al. (2017b). Causes, consequences, and policy responses to the migration of health workers: Key findings from India. *Human Resources for Health, 15*(1), 28.

Warf, B., & Arias, S. (2008). *The spatial turn: Interdisciplinary perspectives*. Milton Park: Routledge.

Watson, A., & Ward, J. (2013). Creating the right 'vibe': Emotional labour and musical performance in the recording studio. *Environment and Planning A, 45*(12), 2904–2918.

Watson, D. B., Murtagh, M. J., Lally, J. E., Thomson, R. G., & McPhail, S. (2007). Flexible therapeutic landscapes of labour and the place of pain relief. *Health & Place, 13*(4), 865–876.

Webster, J., & Kristjanson, L. J. (2002). But isn't it depressing? The vitality of palliative care. *Journal of Palliative Care, 18*(1), 15.

Wei, L. (2004). Using a GIS-floating catchment method to assess areas with shortage of physicians. *Health & Place, 10*(1), 1–11.

West, E., & Barron, D. N. (2005). Social and geographical boundaries around senior nurse and physician leaders: An application of social network analysis. *Canadian Journal of Nursing Research, 37*(3), 132–149.

Whitehead, D. (2001). Health education, behavioural change and social psychology: Nursing's contribution to health promotion? *Journal of Advanced Nursing, 34*(6), 822–832.

Whitelaw, S., Baxendale, A., Bryce, C., Machardy, L., Young, I., & Witney, E. (2001). 'Settings' based health promotion: A review. *Health Promotion International, 16*(4), 339–353.

Wiles, J. (2003). Daily geographies of caregivers: Mobility, routine, scale. *Social Science & Medicine, 57*(7), 1307–1325.

Wiles, J. (2005). Conceptualizing place in the care of older people: The contributions of geographical gerontology. *Journal of Clinical Nursing, 14*, 100–108.

Wiles, J. L., Leibing, A., Guberman, N., Reeve, J., & Allen, R. E. (2012). The meaning of "aging in place" to older people. *The Gerontologist, 52*(3), 357–366.

Williams, P. M., & Shavlik, G. (1979). Geographic patterns and demographic correlates of paramedic runs in San Bernardino, 1977. *Social Science & Medicine. Part D: Medical Geography, 13*(4), 273–279.

Wills, J., Cumbers, A., & Berndt, C. (2000). The workplace at the millennium: New geographies of employment. *Environment and Planning A, 32*, 1523–1528.

Wilson, M. (2002). Making nursing visible? Gender, technology and the care plan as script. *Information Technology & People, 15*(2), 139–158.

Worth, N. (2016). Who we are at work: Millennial women, everyday inequalities and insecure work. *Gender, Place & Culture, 23*(9), 1302–1314.

Yantzi, N. M., & Skinner, M. W. (2009). Providers of care in the home: Sustainable partners in primary health care? In *Primary health care: People, practice, place* (pp. 221–236). London: Ashgate.

York, D. (2008). Medical tourism: The trend toward outsourcing medical procedures to foreign countries. *Journal of Continuing Education in the Health Professions, 28*(2), 99–102.

Zitzelsberger, H., McKeever, P., Peter, E., Chambon, A., Morgan, K. P., & Spalding, K. (2014). Doing 'technological time' in a pediatric hemodialysis unit: An ethnography of children. *Health & Place, 27*, 112–119.

Part I
Introductions

Chapter 3
Case Study I: Hospital-Based Multidisciplinary Work—Institutional Emotional Geographies

Emma Rowland

Abstract This chapter is a case study based on Emma Rowland's original empirical work that showcases some contemporary geographies in professional practice. It presents data derived from ethnographic observations in four hospital wards and interviews with professionals working in them in multiple work roles. The aim is to articulate the character and function of 'emotional geographies' in hospital-based/ward work. Emerging themes include geographies of: emotional detachment, contested emotion, emotional attachment, emotional containment, and communicating emotions.

Introduction

As reviewed in Chap. 2, health professional research has certainly explored the importance of space to place to health care practice through several theoretical lenses. This case study and chapter explores the function of emotions in intimate care relationships between health professionals and their patients in different hospital spaces (Rowland 2014). Doing this frames this inquiry in the field of 'emotional geographies' (Anderson and Smith 2001; Davidson et al. 2012; Bondi 2016)—an empirical yet theoretically rich field, which runs the gamut from humanism, feminism, to poststructuralist and non-representational theory. A field which, Davidson and Milligan (2004) suggest, explores how emotions are practiced and felt across space and in place: how emotions are important in themselves and how they motivate or restrict action (a basis for meaning and identity and also oppression and empowerment); how emotions are critical to our 'being-in-the-world' and help to define our sense of place and self (how places provide and/or draw out emotions; and how emotions create place).

G. J. Andrews et al., *Place and Professional Practice*, Global Perspectives on Health Geography, https://doi.org/10.1007/978-3-030-64179-5_3

Emotions

The subject of emotions unsurprisingly arises frequently across health professional research. Much of it describes, after Goffman (1990), how emotions are performed, controlled and given meaning in health care through social and cultural etiquettes—or "feeling rules". These are thought to control people's outward expressions of emotion, which may be different from what is felt internally. In health care contexts, health professionals perform particular emotional selves to colleagues, patients and relatives. They present an acceptable professional self and suppress and control their private thoughts and emotions. This, however, is hard work, and their performance can easily slip exposing unacceptable emotional displays. Indeed, within the traditional biomedical model, health professionals are expected to present 'respectable emotions' (Milligan 2005, p. 2109) therefore displaying emotionally detached or professional behaviors in front of their patients. Creating physical distance is encouraged to increase the emotional detachment between health professionals and their patients. Health professionals, therefore, should only engage in physical interactions with patients for medical examinations and avoid touching patients to show empathy or give reassurance. In this model, emotional detachment supports objective clinical care by preventing emotions from contaminating the decision-making process (Allan 2001a, b). Furthermore, emotionally detached care behaviors are perceived to enhance health professionals' emotional and psychological well-being by allowing them to 'switch off' emotionally thus preventing fatigue and burn-out (James and Huffington 2004; Mackintosh 2007).

The practice of emotionally detached professional care has been explored in the most emotionally challenging clinical spaces. Exemplary research includes that about nurses who work in 'deathscapes' (Maddrell and Sidaway 2010; Thompson 2012), i.e. those spatialities of care in which there is a high risk of mortality and health professionals are confronted by death on a regular basis, such as Care of the Elderly and Intensive Care Units (ICUs) (Greaves 1994; Mann 2005; McCreight 2005; Mackintosh 2007). Delivering emotionally detached nursing care requires health professionals to construct defense mechanisms, which protect them from the guilt, anxiety and stress of providing care to the sick and dying (Menzies 1970; Smith 1992; Curtis et al. 2012).

Research suggests that health professionals have become disillusioned by the traditional biomedical model of care and are increasingly frustrated by the resulting lack of emotional engagement with their patients (Bolton 2001; Erickson and Grove 2007, 2008). In response, healthcare professionals are becoming more 'emotionally aware' (Allan 2001a, b). This has resulted in a paradigm shift to a more contemporary model of care in which emotions are brought more acutely into focus through careful emotion management, emotional labour and emotion work.

Whilst there is a plethora of research focusing on the emotional labour of nurses, there is a dearth of research that provides insights into the day-to-day emotional management by other health professionals, notably doctors, other allied health professionals and hospital managers (Larson and Yao 2005; Erickson and Grove 2008).

In addressing this knowledge gap, whilst still being attentive to nursing geographies, this chapter draws on empirical ethnographic data to illuminate the emotional care work of health professionals working in the UK's National Health Service (NHS).

Methods

Methodologically underpinned by (sensory) ethnography, this study sought to understand the emotional world of health professionals through inter-subjective, inter-corporeal, embodied and emplaced knowledge (Porteous 1985; Simmel 1997; Pink 2009, 2015). A sensory ethnographic approach allowed the researcher to employ all their bodily senses (sight, smell, taste, touch, aural and visceral) to capture and analyse participants' emotions, making their body an essential data collection tool (Moss and Dyck 1996, 1999; Parr 1998a, b). Personal sensory experiences in the field bore witness to, and enabled the researcher to explore, the emotional and embodied worlds from an emic perspective (Longhurst 1997; Martin 2002; Edvardsson and Street 2007).

Qualitative data were collected from three NHS Primary Care Trusts (PCTs) in England. Within these three Trusts the research was located in five hospitals (Broadwater Hospital,[1] Helios General Hospital, St Joseph's Community Hospital, Royal Alexandra University Hospital and Royal Victoria Infirmary). Data were collected in different medical departments: Cardiology, Care of the elderly, the Emergency department (ED), General Medicine and Obstetrics and Gynaecology.

Ethnographic methods included: non-participant static observations (Madden 2017), participant observation including shadowing (Czarniawska 2008; Bartkowiak and Sappey 2012) and semi-structured and storytelling interviews (Hansen and Kahnweiler 1993; Kendall et al. 2009). Static non-participant ethnographic observations (90 h) took place on hospital wards, in clinics or in waiting rooms with 20 nursing staff. On the hospital wards, observations were predominately conducted from the nurses' station. From this vantage point all patient care activity could be witnessed without being intrusive to the delivery of care and/or prevented patients from feeling uncomfortable due to the researcher's voyeuristic gaze. The nurses' station was a space, hidden from the patients' gaze, where nurses conducted logistical and care management tasks such as filling-in patient records, preparing drugs, organising rotas, completing paperwork and interacting with other health professionals. It was also a space in which health professionals engaged in back region behaviours, such as talking in lowered tones about "difficult" patients, exchanging patient care stories, gossiping about other health professionals and discussing personal issues. This space was therefore invaluable for collecting emotional narratives and to gaining insight into the emotion management of ward nurses.

[1] In presenting this research the hospitals and the health professionals, middle managers and patients within them have all been given pseudonyms to protect their identity.

Ten health professionals were shadowed (200 h). Shadowing involved following the health professional for the duration of their shift (10–12 h) as they moved between patients and across the hospital (McDonald 2005; Neyland 2007), enabling a greater embodied understanding of health professionals' emotional worlds (Pink 2009, p. 76). Throughout all ethnographic observations, ethnographic interviews took place to clarify behaviours, events and interactions, giving further understanding and meaning to observations.

Whilst experiencing visceral reactions and emotions in the field, the researcher was challenged in translating these feelings into words and understanding how they may correlate with their participants' emotions (Rowles 1978; McCormack 2005; Parr et al. 2005). Reflexivity was crucial in aiding emotional interpretation as it allowed the researcher to think about their own emotional responses regarding what they witnessed. The researcher's own experiences and feelings were then used to interrogate how their participants may be feeling. Furthermore, a 'research alliance' between the participant and the researcher was essential (Pile 1991, p. 461) as this enabled participants' deep embedded (unconscious) emotions to be projected towards and transferred onto the researcher's body allowing the researcher to experience and interpret their participants' emotions (Pile 1991; Laurier and Parr 1999). Counter-transference was used to expose the researcher's own feelings and emotional responses to participant, allowing them to think reflexively about the interpretation in a feedback loop (Pile 1991; Widdowfield 2000; Bondi 2005a). Finally, to enhance the translational accuracy of participants' emotions, the researcher became embedded in the social practices of those observed, thus making the transition from an 'outsider' to an 'insider'.

Further to the observations, as suggested semi-structures interviews and storytelling interviews (n = 28) were conducted to gather a range of experiences and emotional narratives from a variety of health professionals. Interviews were conducted in the hospital and lasted on average 60 min. They were audio recorded using an encrypted digital Dictaphone and transcribed verbatim. Hand-written notes were taken to sign-post follow-up questions emerging from the participants' narratives and to record participants' body language or facial expressions.

All qualitative data were inputted into ATLAS Ti 6.2 for data management (Friese 2019) and analysed using thematic analysis (Boyatzis 1998; Braun et al. 2014, 2019). Data analysis produced four themes: emotional detachment, contested emotions, emotional attachment and emotional containment, which will structure the remainder of the chapter.

Emotional Detachment

Different spatialities within the hospital (i.e. different departments and wards) have an impact on the emotional demands of health professionals working within them, resulting in the performance and functionality of different emotion management techniques. Spatalities of care that encouraged emotionally detached care behaviours were those such as the ED, which were defined as temporally 'temporally rich' (Urry 2005), meaning that they are busy and fast paced with a high patient turnover. Other spatialities of care that encouraged emotionally detached care behaviours included 'taskscapes' such as Intensive Care Units (Liaschenko et al. 2011) where health professionals focused more on clinical tasks or medical equipment than providing care to the patient and 'deathscapes' (Maddrell and Sidaway 2010), such as palliative care and Care of the Elderly in which health professionals frequently encountered death. In these care spaces, health professionals were often encouraged to manage their emotions in a way which produced both physical and emotional distance between themselves and their care recipients.

Illustrating the role and function of emotional detachment, the following extract with respiratory consultant Mr. Robert White on a Care of the Elderly ward demonstrates how the traditional medical model of care serves to protect health professionals from emotionally challenging tasks and decision making.

> *... The patient's room is dimly lit and has a strong smell of faeces. As we enter the patient's daughter is trying to encourage her mother to drink a protein shake. She moves away from her mother and stands by her father's side in the top right-hand corner of the room. Louise (registrar) stands by the patient's daughter and husband hugging a red folder containing the patient's medical notes. The patient has a vacant haunted expression. She is thin and frail and her face is gaunt accentuated by her unkempt hair. Robert loudly introduces himself. The patient doesn't respond. Robert tries again. He elevates her arm by taking her wrist and begins to take her pulse she moves her head towards him, stimulated by his touch. Louise comments that the patient seems to be more responsive than this morning...*
>
> *Robert asks the patient if he can listen to her chest and she slowly nods...Louise helps to sit the patient up, resting the patient's body on her forearm as Robert places the stethoscope on the patient's back and then to her chest. The husband holds his wife's hand throughout and then helps Louise lay her back down. Robert informs Louise that he heard 'some crackles' and she draws some x's on the patient notes over a rudimentary picture of a pair of lungs...*
>
> *Robert covers the patient up and then turns to the daughter asking, 'how is she today?' The daughter tells him that she is responding more than she did this morning but she has been refusing food and drink...Robert comments that the patient has good skin condition and a good color in her mouth so is not worried about the water intake, but he is worried about the amount of food she is consuming. Robert talks directly to the daughter stating that her mother is coming to the natural end to her life and therefore she is 'withdrawing from life' and that although they will feed her through a tube she should prepare herself and her family for the inevitable. The daughter says that she understands and turns to her father and says 'Dad, do you understand what the doctor is saying?' The husband paces up and down the side of his wife's bed looking lost and agitated...*
>
> *Robert asks 'how is the family coping?' The daughter says that her father is finding it very difficult because he is her main carer. She has also taken time off work so she can support her father. As she speaks, she becomes tearful. Robert listens and occasionally asks questions. The husband continues to pace the room occasionally touching his wife's head and face. Falling silent the daughter watches her father.*
>
> *In the silence Robert begins to talk about the deterioration of her mother and what they will be able to do for her. The daughter says 'you're talking about DNR aren't you?' Robert confirms that he is...The daughter calls her father over and begins explaining what DNR means. He looks at his daughter, looking lost and scared and she tells him that it is his decision but she thinks that the doctors should not resuscitate her mother because she will be resuscitated to the same state. Robert corrects... [she] will be 'worse' because she will be a*

lot weaker. The daughter says that she will have to talk through this with her father... Robert nods and leaves the room, immediately signing the DNR form and thrusts it into Louise's hands.

Choosing between letting patients die or performing aggressive and invasive treatments often leaves health professionals with a "frightening anxiety about playing God" (Moylan and Jureidini 1994, p. 232). Robert's behavior is therefore perhaps typical of health professionals working in 'deathscapes' (Brown 2003; Maddrell and Sidaway 2010; Liaschenko et al. 2011). Health professionals who encounter death on a frequent basis often use workplace feeling rules to construct an emotional buffer to protect themselves from the emotionally challenging environment (Lief 1963; Fabricius 1991; Allan 2001a, b). This emotional buffer or 'defense mechanism' (Lief 1963; Menzies 1970) also prevents health professionals from becoming consumed by the emotional and affective under-currents that the patient and their relatives are immersed in. Emotional detachment in this way is therefore considered to support health professionals in making better care management decisions, as their cognitive thought processes are not contaminated by emotions, which could lead to medical mistakes (Allen and Kleinman 1989). In this extract, we see that Robert does not want his decision to resuscitate his patient to be motivated by emotion. He wants to separate his emotions from rational, objective thought, drawing on evidence-based practice to make a sound clinical judgement. Robert states that a decision motivated by emotion could be detrimental, causing patients unnecessary pain and extending suffering:

...you learn to distance yourself from the patient otherwise it would be detrimental to your practice...you need a clear head

Robert describes becoming estranged from his emotions through emotion management. In emotionally detaching from their emotions, health professionals can become alienated from their own and their patient's emotions, allowing them to switch off from the emotional consequences of their clinical decision making (Allen and Kleinman 1989; Greaves 1994; Mackintosh 2007). However, whilst Robert asserts that he has become hardened to the impacts of his decision-making, the speed in which the do-not-resuscitate (DNR) form was signed and handed back to his registrar intimates that he may harbor some internal guilt or anxiety. It is here that we gain some insight to the performativity of emotion management as we witness the interplay between Robert's bodily interior and exterior boundaries as he suppresses and contains his emotions.

Further to the DNR form, Robert uses workplace time directives as a defense mechanism to emotionally detach from his patient (Mackintosh 2007). NHS time directives (workplace rules and regulations) restrict the time that health professionals are able to interact with their patients. Time directives compartmentalize health professionals' time into task orientated activities such as taking diagnostics, monitoring, handing out medications, bathing, dressing, mobilizing, etc. However, they do not allocate time for health professionals to build inter-personal relationships with their patients (Smyth 2013). Emotionally detached care behaviors are therefore encouraged by the organization and any guilt experienced by health professionals for not establishing relationships with their patients is therefore absorbed and should be consumed by the NHS, thus reducing health professionals' emotional burden. In highlighting this point, Robert is witnessed only engaging with the patient to seek consent to perform clinical tasks.

Whilst emotionally detached care behaviors have allowed Robert to contain his 'disruptive emotions' and maintain a bounded sense of his professional self towards the patient, Robert demonstrates 'emotional awareness' towards the patient's daughter (Allan 2001a, b; Mark 2005). This illustrates the emotional complexities of care work and emphasizes that health professionals are skilled emotion managers who are able to create emotional detachments and attachments with different patients/relatives in the same spatio-temporal environment. In an ethnographic interview following this observation, Robert illustrates his emotional awareness, commenting:

> *Did you see the anguish in the daughter's eyes....the isolation....loss and tenderness in the father...they are the ones that need my help*

Robert is able to sense, feel and understand the relatives' emotional strains through reading their body language and facial expressions, demonstrating that emotions are both inter-relational and an embodied experience. Robert disclosed in an ethnographic interview that his own experience as a relative in a similar situation with his own family have enabled him to connect with the relatives on an emotional level by drawing on his 'emotional memories' (Hochschild 1979). He is therefore able to support the patient's relatives through this emotionally challenging time, helping them to come to terms with and accept the patient's imminent death through deep acted emotion management (Hepworth 2007).

Interestingly, however, whilst Robert demonstrates an emotional awareness of the relatives' emotions, he only offers an emotionally attached care relationship with the patient's daughter. Emotionally detached care behaviors continue to be performed towards the patient's husband. It could be considered that Robert may have deliberately chosen to become emotionally connected to the daughter, because she appeared to be more capable of managing her emotions. Her father's bodily deportment and anxious pacing clearly indicated his heightened 'death anxiety' at the prospect of losing his wife. By directing communication through the daughter, Robert's emotional performance may be less labored:

> *[I need to] judge the most suitable way to tell relatives, and adapt my approach to make it as painless as possible. However, yesterday was easy for me because the daughter was intelligent and knew all about DNR so I didn't really have to go into any detail as she understood and then could explain it to the father.*

It is here that we see Robert relinquishing his responsibility to deliver bad news to the patient's husband by using the daughter as an intermediate interlocutor for the exchange. Using the daughter in this way, Robert distributes the emotional burden reducing the intensity of his emotional labor by distancing himself from the husband's reaction and protecting his own emotional well-being (Aldridge 1994).

Contested Emotions

Robert's account demonstrates not only how emotional detachment is used as a defense mechanism to reduce professionals' anxiety and guilt in delivering patient care through workplace rules or task orientated behaviors. It also highlights the complexity of emotional care work within a singular spatial-temporal environment. Further, this illuminates the complexity of emotion management by highlighting the emotional struggles faced by health professionals as they try to reconcile workplace rules with their own personal feeling rules. For some, a health professional's workplace rules were unhelpful, and they did not reduce their emotional burden. Instead, workplace rules increased health professionals' sense of guilt and anxiety by preventing the delivery of emotionally attached care which they desired to give. In response, some health professionals resisted the constraints set by organizational rules to deliver emotionally attached care behaviors. These health professionals delivered emotional care work within the liminal space between emotionally attached and detached care, resulting in *geographies of contested emotion* (Bondi 2005b).

The desire to become emotionally attached to patients and to talk to them on an emotional and social level has become increasingly important to health professionals who are becoming burnt-out and dissatisfied with the emotionally disconnected and disembodied medical model of care (Aldridge 1994; Bolton 2001). The extract below is taken from an interview with Mr. Alexander Hopper, Consultant Cardiologist, who emphasizes his desire to become emotionally connected to his patients to increase patient satisfaction and his own job satisfaction (Redinbaugh et al. 2003). This desire leads to an internal emotional struggle as personal, organizational and/or professional rules collide. Alexander's interview therefore raises an important exploratory point for nursing and emotional geographies of care.

ER *… how do you personally try to provide good patient care?*

AH *Hmm, ((2)) well that's a good question. Erm, ((laughs)) well, I think in terms of care, one needs to listen to people and their problems. If you're not prepared to give people the time to tell you their story you may miss things, erm, experience helps to some extent…I think you've almost got to create [time], erm… I mean sometimes it seems I'm under a bit more pressure, I'm doing a ward round tomorrow at 8am and another one on Friday and the number of people coming in overnight may vary from 3 to 12. So, if I have 12 to see and then the rest of the ward that's already a lot of patients and you feel under more time pressure. But I think you've just got to spend time really and not feel rushed, because if you rush it you're going to miss things and it does take time… we tend to forget that they are individuals in front of us, they are mothers, fathers, husbands, wives, children of people like us and they deserve time spent on them.*

ER *Do you see them as people when you treat them?*

AH *…Yes, because they're always different and they're always people. I sit in with the juniors on Monday and a student consulted with a patient and I said to him, and this man's retired and I said to him, what did he do? You know? What was his role, he's 75 now, and he said I don't know. I said well why don't you know? You're meant to be looking at this person as a person, not just some sick body who just came to hospital. You need a holistic view of the individual and that's why we talk to them about their background, and it's trying* *to create a complete picture of somebody. It's difficult to do that in 10, 15 minutes, but we try.*

The fast paced, temporally rich 'taskscape' of this cardiology ward is a source of anxiety for Alexander because he is unable to spend the amount of time with his patients that he feels is sufficient to gain holistic (emotional, social and physical) knowledge of his patients' needs. He believes that if the hospital ward environment was temporally poorer (Urry 2005, p. 80) it would allow him more time to establish emotional care relationships with his patients, which would lead to more informed decisions about his patients' care pathways. In contrast to Robert, Alexander believes that emotionally attached care assists clinical decision-making rather than inhibits it (Mumby and Putnam 1992; Mayall 2010).

Aligning with Alexander's feelings of anxiety roused by the spatio-temporal constraints, Bolton (2001) reveals that many health professionals feel increasingly pressured, frustrated and anxious about the amount of time that they are able to spend with their patients due to increasing patient numbers and decreased human resources (Erickson and Grove 2007). In response to the temporally rich environment, health professionals, particularly nurses, are therefore becoming more creative about the time they give to their patients (Bolton 2000, 2001). Alexander, like many other health professionals observed, has learnt to be more efficient with his time allowing him to provide more holistic patient-centered care to his patients and reducing his guilt and anxiety. However, for Alexander, temporal creativity has not led to total personal satisfaction, leaving him feeling frustrated and underwhelmed by the care he can deliver.

During his interview Alexander appeared anxious about his "failings" to establish emotionally attached care relationships with his patients, although these struggles were not visible during my observations of his care work. Instead, Alexander was observed to have very good rapport with his patients and emotionally attached care behaviors were witnessed through his emotionally embodied practice as illustrated below:

The team head to a cardiac step down ward. John (doctor F2), summarizes the patient as he glances over the patient's file. Alexander comments, "ah, I remember him from a previous admission" ...The team visits the patient's bedside. On entering the bay the patient calls out 'Dr Hopper! How lovely to see you!' and holds out his hand to shake it. Alexander rapidly walks to the patient beaming and shakes his hand and then sits on the patient's bed 'How are you old chap?'

Alexander *has* established an emotional attachment with his patient, with their bodily proximity and informality, emphasizing their emotional care relationship. Alexander actively encourages bodily proximity by sitting on the patient's bed and sharing in the patient's personal space. Neither Alexander nor the patient are perturbed by the gesture of physical closeness suggesting that the emotionally connected relationship is reciprocal and desired (Roter et al. 2006; Evans et al. 2009). Alexander's act also removes doctor-patient bounded hierarchies, breaking down the power relationships that are created between the health professionals and their care recipients during the delivery of care.

This interaction not only demonstrates that Alexander has become connected to his patient through bodily proximity but also through surface acted emotional labor. Whilst it is evident that the patient believes Alexander's emotional performance, perhaps leading to an enhanced patient care experience, Alexander is unable to convince himself of the emotional performance, leaving him feeling distressed and emotionally ambivalent. This is emphasized below:

> *At the end of the ward round Alexander says that he would like to talk through what I have observed. Alexander makes it clear that he feels very guilty and worried about the time that he spends with his patients. He tells me that he wishes that he could spend more time with his patients and that, if he was a relative, he would like the doctor to spend a lot more time with his loved one than he gets to spend with his patients.*
>
> *Alexander continues that he would like to provide more holistic treatment to his patients. However, talking to patients to find out their social circumstances and well-being takes time, time that he simply does not have. He complains that he finds himself only concentrating on the medicine and…the hard facts… He seems worried that in doing this he could potentially miss something important. This predicament leaves Alexander feeling both stressed and guilty.*

Alexander blames his inability to establish an emotional connection with his patient on the ward's spatio-temporal constraints and shifting workplace and professionals' feeling rules, which only allow him to collect only the 'hard facts'. Alexander, therefore, only performs surface acted emotional labor (Milligan 2005). Alexander's internal emotional battle demonstrates that workplace feeling / emotional rules are an "integral feature of emotional geographies because they exist as essential cultural, [organizational and professional] guides or bridges through the complex and difficult pathways" of care work (Hepworth 2007, p. 189). We have witnessed here how workplace feeling rules govern subjective emotions. In addition, we have seen that they may be challenged and contested by personal feeling rules about how health professionals should perform care work. This therefore emphasizes the challenges in achieving and delivering emotional attachment in carescapes.

Emotional Attachment

Shifting focus, we now move on to explore the performance of emotionally attached care work. The extract below is taken from an interview with Judith King, Matron of the General Medicine ward at St Joseph's Community Hospital. It illustrates how hospital wards that are 'temporally poor' (Urry 2000, 2005) can facilitate emotional connections between health professionals and their patients. Temporally poor hospital wards, such as this General Medicine ward, are characterized by slow patient turn-over, reduced patient care activities and low patient mortality rates. Patients on this General Medicine ward require limited or no medical intervention and may be waiting for social packages before they can be discharged. Ethnographic notes on Judith King, Matron, raises an important point on ownership and belonging, they reading:

Judith heads to Wordsworth Ward 2 where we stand behind the nurses' station. Judith begins by checking the white board and then looks through the files of the new patients that were admitted overnight. Judith (looking at the white board) talks me through the different types of patients that they have on the ward and then begins reading through the new patients' notes to get to know her patients. As Judith reads through the files I look at the board and comment that some of the patients have been on the ward a very long time. Judith agrees and informs me it is because many of the patients are elderly with complex medical conditions and therefore their admission is extended. I note that one patient in particular has been here for three months and ask whether she finds herself building relationships with any of the patients. Judith looks up from the file and at the white board says, 'Oh, that's Gloria, mmm has she been here 3 months? Let me check her notes'. Judith leaves the file that she is reading and picks up Gloria's file and looks at the admissions page and then states (sounding surprised) 'three months is correct'... Judith admits that she does find herself getting attached to her patients, especially if they have been here as long as Gloria. She continues that Gloria 'should be elsewhere, but she is ours' explaining that Gloria is really sweet and all the nurses really like her. It would be quite strange if Gloria was no longer on the ward as 'she is a returning patient and it is always nice to have her back'.

Judith continues to talk about Gloria and I discover that her husband was also in the hospital, but unfortunately had died during his last admission. Following the death of her husband Gloria had come in to see all the nurses with her grandchildren because she had wanted to see some 'friendly faces'... Judith pauses and stares at the board, sighs and then says (sounding disappointed) that they had only managed to make Gloria well for 2 months out of the year and for the other 10 months she has been constantly in and out of the hospital. Although clearly defeated by illness and old age, Judith believes that she and the nurses 'have made a real difference to her life' and hopes that Gloria will get better again as 'she is such a lovely character'.

Judith explains that she has become emotionally attached to Gloria through the ward's temporally poor environment and Gloria's status as a returning long stay patient, which has meant that they have had time to get to know each other on an interpersonal level. This has led Judith to experience a sense of ownership over Gloria, which was emphasized by Judith's comment that "she should be elsewhere but she is ours". Judith's emotional connectivity with Gloria is witnessed not only through Judith's description of their relationship, but with her emotional talk surrounding her relatives; her husband and her grandchildren.

Setting Gloria within her familial context humanizes her, allowing intimate emotional connections to flourish, as patients are regarded as people and not a set of disembodied, dismembered body parts. These emotional connections benefit health professionals through meaningful therapeutic relationships (Suchman and Matthews 1988; Mattingly 1998; Larson and Yao 2005). In talking about Gloria's increasing ill-health, however, Judith's tone begins to change and there is an indication that Gloria's ill-health is causing the nursing team's emotions to become increasingly labored. A change in emotional labor may result in a change in the emotional care relationship offered to Gloria. This relationship may shift between one that is emotionally attached to an increasingly emotionally detached care relationship as Gloria's risk of death increases. The notion of 'ownership' may also shift as the patient's risk of mortality increases.

Whilst Judith displayed a sense of ownership over Gloria, Judith perceives that Gloria reciprocated these emotions by feeling an emotional attachment to the ward and the health professionals working within it. This may have resulted in the sense of belonging. Gloria's sense of belonging was established and strengthened through her long admission, which enabled her to become emotionally attached to the nurses on the ward. Gloria's sense of belonging or "emotional affiliation" (Wood and Waite 2011, p. 201) to the ward and the nursing team was illuminated through Judith's narrative about Gloria's return to the ward following the death of her husband in which she returned not as a patient but to "see some friendly faces". Gloria therefore longed and yearned for some support and "emotional refuge" (Fields 2011, p. 258) in a place in which she felt safe and secure, at a time when she felt most vulnerable, isolated and insecure. The hospital ward therefore has an affective and emotional identification (Caluya 2011) for Gloria as a space of emotional care.

Gloria's sense of belonging outside of the patient care boundaries demonstrates a wider affective relationship between herself and the nurses. To provide care to Gloria outside of her capacity as a patient, the nursing team must employ emotion work to support her. By providing care through emotion work, Judith's nurses give a part of their emotional selves to Gloria (Bolton 2000; Erickson and Grove 2008; Deery and Fisher 2010), which strengthens their emotional bond. This is emphasized further when Judith is observed to bend the workplace rules to allow Gloria to stay on the ward, even though she should be relocated. In an ethnographic shadowing with Consultant Geriatrician Madeline Marsden, workplace rules were also bent to allow two elderly patients to remain in her care on the elderly ward. Ethnographic notes on Madeline Marsden, consultant geriatrician, read:

> *...'Next patient?' Sam (F2, doctor) collects three patient files from the trolley and tells Madeline that they are moving along to Flo and Doris. Madeline turns to me and tells me that these patients have been on the ward for a long time as they await social packages. She informs me that the two ladies have become inseparable since they have been admitted and have enjoyed their stay because they have spent all their time chatting and gossiping. She continues that she needs to transfer them to another ward so they can free up some bed space, but they are reluctant to move one without the other, because they have become such great friends, which has meant that for now they will remain on the ward.... Madeline states that they will 'certainly be missed' by the nursing staff and herself, as they have great characters.*

In both Madeline and Judith's ethnographic examples, it is made clear that these patients are logistically 'out of place'. A sense of ownership or an affective affiliation, however, has allowed these patients to remain on the wards. The emotional attachments were not only established through the length of time that these patients had spent on the ward but also through their characters. Whilst Zappi and Epstein (2000) personal correspondence stated that "caring is a charitable act and occurs regardless of liking the patient", it is clear from the extracts that health professionals do not always provide equitable care and that care can be influenced by whether patients are well liked or less challenging than others. Health professionals have the capacity to bestow 'emotional gifts' to favored patients. In these extracts, both Madeline and Judith demonstrate altruistically motivated behaviors by allowing their patients to stay on the ward. However, because "spontaneous goodwill...is [often] an insufficient explanation for emotional gifts" (Cohen 2010, p. 213), it could also be suggested that these gifts could have been given with ulterior motives (Erickson 2005; Erickson and Grove 2008). It could be considered, for example, that these patients have been allowed to stay on the ward because they are "good patients", requiring minimal emotional labor from health professionals and therefore reducing emotional burn-out and fatigue.

To further this point, the emotional demands of caring for challenging patients were observed on numerous occasions, and health professionals were witnessed bending organizational rules to protect themselves from burn-out, fatigue and poor job satisfaction. During static ethnographic observations on Wordsworth Ward 1, for example, a General Medicine ward at St Joseph's Community Hospital, nurses were overheard stating that they did not want a patient to be moved as this would result in another patient occupying the bed. This incoming patient was a well-known returning patient and was deemed both "rude" and "ungrateful" by the nursing staff. To care for this patient required intensive emotional labor, with nurses working hard on their emotional performances to contain negative emotions and display appropriate organizational behaviors such as empathy (Bolton 2001; Hall et al. 2002; Larson and Yao 2005; Roter et al. 2006). The challenges and demands of emotional containment within defined temporal-spatial environments are explored below.

Emotional Containment

Health professionals contain their workplace and personal emotions to perform empathy to patients (Larson and Yao 2005). Empathy has been shown to be "fundamental to caring and enhances the therapeutic potential of health professional / patient relationships" (Larson and Yao 2005, p. 110). Health professionals who display empathetic care are regarded as more effective carers due to the establishment of affective bonds. These bonds strengthen the emotional connectivity between health professionals and patients within the carescape (Larson and Yao 2005; McCreight 2005). This results in "patients being more forthcoming about their symptoms and concerns, thus facilitating medical information gathering, which in turn yields more accurate diagnosis and better patient care" (Halpern 2001; Larson and Yao 2005, p. 110). Lucid narratives taken from a storytelling interview with Dr. Nikhil Chopras, consultant in Obstetrics and Gynaecology, illustrates the importance of empathy in patient care:

> *...I feel comfortable controlling my emotions, so that it doesn't affect patient care. Obviously if you let your negative emotions get in the way then you are an unsafe doctor...period! But as long as you are there, looking after your patients and empathise with them.... It is very important to empathise Emma, I can't tell you how important that is ... If you don't feel for what they feel you will not be sensitive enough then they will not feel it...you know we talked about positive vibes, we talked about all that affecting the patient...you know it is very contagious...*

For health professionals to perform empathy, their bodies need to be receptive to patients' emotional projections so that they may sense and understand how patients are feeling. Nikhil indicates that health professionals are aware of the psychoanalytical process of transference and counter-transference within care interactions and, furthermore, that they are able to use these processes to reassure and support their patients through physical touch and the movement of affects or 'vibes' through the therapeutic space of the hospital ward. Nikhil therefore shows that emotions are not bound within the body, but rather flow back and forth in an affective dialogue to produce an "affective ambiance" (Hubbard 2007, p. 124). The body is therefore a "site of spatial orientation" that allows us to experience and understand the world from others' perspectives through our social skin (Low 2003, p. 13).

The candid narrative that follows tells a harrowing story about the delivery of a still born baby. It powerfully demonstrates the importance of health professionals being emotionally and 'psychologically present' when interacting with their patients (Hirschhorn 1988; Kahn 1992). In addition, this emphasises how both health professionals' and patients' unconscious emotions become entangled within place during care work, especially during emotionally traumatic events (Lewis 2005; McCreight 2005; Kenworthy and Kirkham 2011) where challenging emotions arise (Willis 2009). This extract is important for geographies of nursing because it emphasises that that bodies and emotions and the spaces they inhabit, are inseparable (Duffy et al. 2011):

> ***ER** "Can you give me an example of when you provided good patient care?"*
>
> ***NC** "…I think that …there are a group of patients that… probably need a little more care…not just physically but mentally as well. And one of them was this lady who…came to the labour ward… she told me that she hasn't felt her baby move in the past day and she had had some bleeding… obviously there were some concerns on my behalf for the baby. This was a very wanted IVF pregnancy. On examining her I found that the baby did not have a heart beat…I can only just imagine what that patient must have gone through. I don't think that there could be anything worse than losing a baby, because it can be exceedingly traumatic. It is not only traumatic for her but her relatives; you can see couples completely shattered by the news that I have just given them.*
>
> *Obviously as a human being I do have emotions as well and it wasn't the best of times at all. Erm, sometimes it does affect you, working on the labour ward… it's a huge roller coaster ride -you are delivering someone and there are smiles all the way, on the other hand you… deliver dead ones, which is not the nicest thing to do, yeah, it gives me the creeps just thinking about it.*
>
> *…I did explain everything to her and, as a doctor I need to follow all the procedures. You know I explained it to her that this is by far something that words cannot express… You can imagine how difficult this situation would be and erm, however, if we could make it the slightest bit better, we would try everything in our power you know to make the situation a lot more comfortable. I ensured that she was given a quiet room with a very good midwife, obviously I had my duties towards the rest of the labour ward as well.…We… induced her because we did not want her to go home with a dead baby inside her. It was a very, very painful and stressful scenario…it was horrible… absolutely distressing, exceedingly… distressing because you know the woman is going through all that pain of labour. And labour is by far one of the worst… It is exceedingly painful and I don't know how women take it… Probably because usually it is a good outcome. But in her case it was even worse, she is going through all that pain knowing that at the end of the labour she is giving birth to something that will not survive…It was very, very stressful.*
>
> *So my job was to go in and make sure everything was up to date and that she knew exactly what was happening. I was with her most of the time, but I booked her three-four*

counselling sessions...it is nice to get your emotions out at times. I tried to make it a point that I was present, present as much as I possibly, feasibly could be during her emotional outbursts. I wanted her to have all the support that she needed...I did get attached to the patient, being around her, trying to get her care, even if I was out of hours, which means that when I was not meant to be in the hospital... Some people will argue that, that is a bit over the top, but... I guess that that is medicine for you...It is about lives, and sometimes you have to make that extra effort, to make that person feel that little bit more special.... I thought she would feel a lot more reassured if she saw the same doctor coming in and making sure that she was ok.

Well, you know that when things get worse, they get worse! Ultimately what happened with her was that we tried all the induction of labours, we tried everything possible to get that baby out ...it failed! I can't... tell you how dreadful that scenario was, that she needed a caesarean section to have that baby out....I wanted to make sure that I was present at the caesarean section as well so that she would have a face that she knows...Can you imagine going through a caesarean section, to have a baby...That scar will remind her of her dead baby for the rest of her life!.. While she recovered from her caesarean I saw her every single day on the post-natal ward. Obviously she wasn't quite on the post-natal ward because of course post-natal wards are where women are there with their babies and she didn't have one. So she was in the corner of the labour ward. And I saw her and... made a point not to neglect her, or... her partner because people assume that the only person going through the trauma is the woman... but they are a couple... And he appreciated that quite a lot... They went home. Two weeks later I received a lovely letter, which expressed a lot of gratitude.

Nikhil's monologue illustrates psychoanalytically informed geographies of emotional care work in three ways. First, it demonstrates that, in a medical crisis, health professionals contain their emotions to focus on the parents and their (emotional) trauma. Second, that health professionals communicate emotion. Third, that health professionals have the ability to present emotional gifts to their patients through the performance of their own personal emotions (emotion work).

According to Menzies' research, the majority of hospital work contains a "constant sense of impending crisis" (Menzies 1970, p. 26). This results in an environment which is highly stressful and full of anxiety (Allan and Smith 2005). Whilst Menzies' research focused on nurses, it has been noted that anxiety is exacerbated with increased medical responsibility. Therefore Matrons (Allan and Smith 2005) and Doctors, especially Consultants and Registrars, who have ultimate responsibility for their patients' and medical team's (emotional) behaviors, may experience the greatest anxiety (Huffington and James 2004). To cope with anxiety in a crisis, health professionals must become an "emotional sponge" (Allan and Smith 2005, p. 24) to soak up, contain and manage their own, colleagues' and patients' emotions and "keep them at distance from... their conscious" (Freud 1977; Parr et al. 2005, p. 89).This helps them to perform and display a bodily exterior which is "professionally calm" (Allan and Smith 2005).

In the above context, as the narrative commences, the patient's anxiety is visible. Nikhil attempts to contain his patient's anxiety by conducting medical examinations, which the patient is anticipating (McCreight 2005). Following NHS procedures and conducting task orientated activities, as we have seen previously in this chapter, helps Nikhil to contain his own anxieties. Similar to Robert, Nikhil uses workplace rules and guidelines to focus on other aspects of care work, enabling him to push traumatic emotions to the unconscious, thus helping him to repress and contain his emotions. However, unlike Robert, Nikhil does not display emotionally detached behaviors towards his patient and her husband. Instead, repressing and containing his emotions enables him to be emotionally present.

Nikhil explores his medical responsibility to all of his patient during this crisis. In leaving this patient to fulfil his duty of care to his other patients, Nikhil is able to gain "emotional refuge" (Fields 2011, p. 258) and solace through delivering healthy babies. Engaging in less emotionally strenuous care work bolsters Nikhil's emotional spirit and provides an emotional respite from the emotional burden created by this crisis. Carrying out other care duties, however, induces Nikhil's private emotions (guilt) to enter the public sphere. To satisfy and contain his guilt, Nikhil sacrifices himself to the patient and her husband by being increasingly more visible and "psychologically present" (Klein 1946; Hirschhorn 1988; Kahn 1992, p. 321) by going beyond his "normal" duty of care (workplace rules) and giving a piece of himself (his time) to provide the couple with a "sense of [his emotional] self" (Meier et al. 2001, p. 307). Nikhil is therefore able to move beyond workplace rules "at [his] own discretion…to add something extra to the patient / carers relationship" (Bolton 2000, p. 582).

Aligning with the health professional's emotional containment regarding patients' emotions, it is important for health professionals to understand and communicate these emotions safely. In Nikhil's extract, we witnessed how psychoanalytical processes of transference and counter-transference were used to illustrate how health professionals and patients communicate emotions through unconscious projection. Furthermore, we witnessed how health professionals manage these unconsciously projected emotions (Meier et al. 2001). In this narrative, the patient unconsciously projects and transfers her emotions on to Nikhil's body as she desperately tries to rid herself of the terrors related to her still born baby (Moylan and Jureidini 1994). During this transference, Nikhil experiences the patient's emotions and attempts to manage and contain her "disruptive" and painful emotions (Willis 2009) through counter-transference. However, during traumatic events such as a still-birth, a health professional's unconscious emotions can become entangled making it difficult to provide effective empathetic care.

At several junctures in the monologue, emotional entanglements are witnessed as Nikhil's unconscious emotions surface through emotional slippages, resulting in the audience wondering whose emotions Nikhil is communicating. Nikhil describes the situation as "very painful and stressful" and "horribly distressing". The audience is unclear whether Nikhil is communicating the patient's physical pain of labor and her distress in having to deliver her still born baby or his own emotional pain and stress in having to be responsible for the management of his own, his colleagues' and the patient's emotions during this medical crisis. Nikhil, it seems, is communicating both. Nikhil also states, "it's nice to get your emotions out at times". Whilst Nikhil is referring to the benefits for the patient in talking to a counsellor, he also indicates that talking about emotionally difficult medical situations can also be a cathartic experience for health professionals. This was echoed at the end of the interview when Nikhil expressed his pleasure in taking part in the research, because it had allowed him to talk through situations and emotions that he had not discussed before. Nikhil's admission that he had not communicated this emotional burden to other health professionals also exemplifies the NHS' organizational and professional culture. Talking to colleagues about emotional experiences are still, for some health professionals, regarded as taboo (Bolton 2000, 2001).

Whilst Nikhil's emotional slippages demonstrate the complex emotional entanglements that occur within the emotional topography of the hospital ward, it also provided examples of how health professionals can become interlocutors (communicators) of their patient's emotions. Effective interlocutor health professionals must establish emotional attachments with their patients. They must imagine and visualise the world through their patient's eyes so that they can better understand what their patient is feeling. Health professionals can also become interlocutors of emotion through accessing their own personal experiences and emotional memories. This shows the interplay and complexity of emotion management in care work as both emotional labour and emotion work often combine when health professionals are unable to effectively contain their personal emotions from entering the labour force.

Sometimes the introduction of emotion within the workplace is unconscious; however, there are occasions when health professionals consciously allow their emotional experiences into their place of work to help them contain patient's emotions and to dissipate their own emotional burdens. The introduction of conscious emotion work within the caring professional is regarded as an "emotional gift" (Bolton 2000).

The "death [of a baby is] regarded as a [medical] failure" (Mander 1994; Hunter 2001; McCreight 2005, p. 439). Nurses and doctors who perform terminations of pregnancy, or deliver still born babies, therefore often struggle to cope with the feelings of guilt and anxiety associated with the loss (Lewis 2005; McCreight 2005; Kenworthy and Kirkham 2011). In response to a series of 'medical failures' health professionals such as Nikhil may seek atonement by going beyond the duty of care to provide the patient a wealth of emotional gifts. These gifts might also help health professionals to contain and control patients' negative emotions.

Drawing on Melanie Klein's object relations theory, an interpretation is that gifts are given for reparation to satisfy guilt and sorrow and therefore gift giving is not conducted out of genuine altruism but to satisfy one's emotional needs. Reciprocal gifts are always expected through 'gift-exchange',; for example, Nikhil talks about receiving a thank you card in recognition and gratitude for his emotion work (Hochschild 1983, 2012; Klein 2011). This demonstrates that emotional gifts have exchange value (Hochschild 1983, 2012). Within care relationships, it is important that both health professionals and patients benefit from the gift to protect their emotional and psychological well-being. In seeking atonement for medical failures, Nikhil bestows four emotional gifts onto his patient and her husband. First, he ensures that the couple are given a "good midwife" to look after them. By providing the couple with a good midwife, Nikhil is able to pass on responsibility for the patient in his absence, and thus distributing the emotional strain of caring for this patient across the medical team. The distribution of the emotional strain is also indicated throughout the narrative as Nikhil switches between "I" and "we" when talking about inducing the baby and performing the caesarean, thus subconsciously and psychologically spreading the emotional burden of this crisis.

Second, Nikhil gives the couple a side room off the labour ward so that his patient is physically distant from the other new mothers. This distance prevents the patient and husband from being confronted with the emotional pain of losing their baby. Placing the couple on the neo-natal ward would have taunted the couple leading to greater anxiety, distress and other painful emotions. Furthermore, the side room is used to contain the patient's emotions within a small bounded space preventing their negative emotions from becoming entangled with the "cheerful environment" (Mann and Cowburn 2005, p. 155) or joyous affective ambiance (Hubbard 2007) of the labour and post-natal wards. The containment of the patient's emotions in this side room also makes it easier for Nikhil to manage.

Third, Nikhil provides the couple with counselling sessions so that the patient and her husband can talk through their emotions with a professional. This counsellor will contain and manage the couple's emotional ruptures in an appropriate therapeutic space by absorbing the patient's projected and transferred emotions and managing them through counter-transference, reassurance and empathy. By providing counselling sessions, Nikhil's emotional labour is reduced, and the patients' emotions will be contained and managed.

The final emotional gift provided by Nikhil is that of his time and visibility. Nikhil states that he made a conscious effort to be increasingly more visible and psychologically present with this couple - in a profession where, as already noted, time is very precious (Mattingly 1998, p. 121), Nikhil is seen to be going against the social, cultural and organisational norms by making himself more available to the patient, even coming into hospital on his days off. Reflecting on his behavior, Nikhil states that some of his peers may have criticized him for being too emotionally connected with his patient. However, in showing commitment and professionalism in the face of adversity, Nikhil may have hoped to impress his seniors and achieve career progression or may have wanted to set a good example to his junior colleagues, reinforcing the importance of empathy and compassion in care and thus emphasizing the paradigm shift from the traditional model of care to a new model of care that is more attune to emotions.

Conclusion

This chapter has presented valuable insights into the complex inter-relationships between health professionals, clients and their physical, narrative and emotional proximities in clinical environments (particularly through their different notions of 'belonging'). It emphasised how health care practices are also emotional practices shaped by the nature and temporality of the healthcare environment.

Specifically, the chapter demonstrated how health professionals working in diverse carescapes often decreased bodily proximity and reduced haptic qualities to make sound clinical decisions that are supposedly 'uncontaminated' by emotion.

The idea being, in line with the traditional model of care, that detached behaviours benefit patients by providing respectful and dignified care to patients, especially to those at the end of their life-course. In addition, an idea was that disconnected care supported and protected health professionals from emotional fatigue, anxiety and guilt.

Conversely, this chapter showed how, in some carescapes, health professionals became disillusioned with detached care behaviours leading to geographies of contested emotions. It was clear that in many cases emotionally connected care enhanced health professionals' knowledge and understanding of their patients' emotional world, leading to greater decision-making, care quality, empathy and emotional giving. While reducing health professional's emotional guilt and anxiety and increasing their job satisfaction, emotionally attached care behaviours also enhanced patient's care experiences through emotionally connected care relationships.

References

Aldridge, M. (1994). Unlimited liability: Emotional labor in nursing and social-work. *Journal of Advanced Nursing, 20*(4), 722–728.

Allan, H. (2001a). A 'good enough' nurse: Supporting patients in a fertility unit. *Nursing Inquiry, 8*(1), 51–60.

Allan, H. T. (2001b). Nursing the clinic and managing emotions in a fertility unit: Findings from an ethnographic study. *Nursing Inquiry, 8*, 51–60.

Allan, H. T., & Smith, P. (2005). The introduction of modern matrons and the relevance of emotional labour to understanding their roles: Developing personal authority in clinical leadership. *International Journal of Work Organization and Emotion, 1*(1), 20–34.

Allen, C. S., III, & Kleinman, S. (1989). Managing emotions in medical school: Students' contacts with the living and the dead. *Social Psychology Quarterly, 52*, 56–69.

Anderson, K., & Smith, S. J. (2001). Emotional geographies. *Transactions of the Institute of British Geographers, 26*(1), 7–10.

Bartkowiak, I., & Sappey, J. (2012). The methodological identity of shadowing in social science research. *Qualitative Research Journal, 12*(1), 7–16.

Bolton, S. (2001). Changing faces: Nurses as emotional jugglers. *Sociology of Health & Illness, 23*, 85–100.

Bolton, S. C. (2000). Who cares? Offering emotion work as a 'gift' in the nursing labour process. *Journal of Advanced Nursing, 32*(3), 580–586.

Bondi, L. (2005a). Making connections and thinking through emotions: Between geography and psychotherapy. *Transactions of the Institute of British Geographers, 30*(4), 433–448.

Bondi, L. (2005b). The place of emotions in research: From partitioning emotion and reason to the emotional dynamics of research relationships. In J. Davidson, L. Bondi, & M. Smith (Eds.), *Emotional geographies* (pp. 231–246). Farnham: Ashgate.

Bondi, L. (2016). Emotional geographies. Routledge.

Boyatzis, R. E. (1998). *Transforming qualitative information: Thematic analysis and code development*. Thousand Oaks: Sage.

Braun, V., Clarke, V., Hayfield, N., & Terry, G. (2019). Thematic analysis. In L. Pranee (Ed.), *Handbook of research methods in health social sciences* (pp. 843–860). New York: Springer.

Braun, V., Clarke, V., & Rance, N. (2014). How to use thematic analysis with interview data (process research). In N. P. Moller & A. Vossler (Eds.), *The counselling & psychotherapy research handbook*. Thousand Oaks: Sage.

Brown, M. (2003). Hospice and the spatial paradoxes of terminal care. *Environment and Planning A, 35*(5), 833–851.
Caluya, G. (2011). Domestic belongings: Intimate security and the racial politics of scale. *Emotion, Space and Society, 4*(4), 203–210.
Cohen, R. L. (2010). When it pays to be friendly: Employment relationships and emotional labour in hairstyling. *The Sociological Review, 58*(2), 197–218.
Curtis, K., Horton, K., & Smith, P. (2012). Student nurse socialisation in compassionate practice: A grounded theory study. *Nurse Education Today, 32*(7), 790–795.
Czarniawska, B. (2008). *Shadowing and other techniques for doing fieldwork.* Milton Park: Marston Books.
Davidson, J., & Milligan, C. (2004). *Embodying Emotion, Sensing Space: introducing emotional geographies. Social and Cultural Geography, 5*(4), 523–532.
Davidson, J., Smith, M. M., & Bondi, L. (2012). *Emotional geographies*. Farnham: Ashgate.
Deery, R., & Fisher, P. (2010). 'Switching and swapping faces': Performativity and emotion in midwifery. *International Journal of Work Organisation and Emotion, 3*(3), 270–286.
Duffy, M., Waitt, G., Gorman-Murray, A., & Gibson, C. (2011). Bodily rhythms: Corporeal capacities to engage with festival spaces. *Emotion, Space and Society, 4*(1), 17–24.
Edvardsson, D., & Street, A. (2007). Sense or no-sense: The nurse as embodied ethnographer. *International Journal of Nursing Practice, 13*(1), 24–32.
Erickson, R., & Grove, W. (2007). Why emotions matter: Age, agitation, and burnout among registered nurses. *Online Journal of Issues in Nursing, 13*(1), 1–13.
Erickson, R. J. (2005). Why emotion work matters: Sex, gender, and the division of household labor. *Journal of Marriage and Family, 67*(2), 337–351.
Erickson, R. J., & Grove, W. J. (2008). Emotional labor and health care. *Sociology Compass, 2*(2), 704–733.
Evans, A. M., Pereira, D. A., & Parker, J. M. (2009). Discourses of anxiety and transference in nursing practice: The subject of knowledge. *Nursing Inquiry, 16*(3), 251–260.
Fabricius, J. (1991). Running on the spot or can nursing really change. *Psychoanalytic Psychotherapy, 5*(2), 97–108.
Fields, D. (2011). Emotional refuge? Dynamics of place and belonging among formerly homeless individuals with mental illness. *Emotion, Space and Society, 4*(4), 258–267.
Freud, S. (1977). *Introductory lectures on psychoanalysis.* New York: W.W. Norton & Company.
Friese, S. (2019). *Qualitative data analysis with ATLAS.* Thousand Oaks: Sage.
Goffman, E. (1990). *The presentation of self in everyday life.* London: Penguin.
Greaves, J. (1994). Normal and abnormal grief reactions: Midwifery care after stillbirth. *British Journal of Midwifery, 2*(2), 61–66.
Hall, J. A., Horgan, T. G., Stein, T. S., & Roter, D. L. (2002). Liking in the physician–patient relationship. *Patient Education and Counseling, 48*(1), 69–77.
Halpern, J. (2001). *From detached concern to empathy: Humanizing medical practice.* Oxford: Oxford University Press.
Hansen, C. D., & Kahnweiler, W. M. (1993). Storytelling: An instrument for understanding the dynamics of corporate relationships. *Human Relations, 46*(12), 1391–1409.
Hepworth, M. (2007). Ageing and the emotions. Framing old age in victorian painting. In *Emotional geographies* (pp. 177–190). Aldershot: Ashgate.
Hirschhorn, L. (1988). *The workplace within: Psychodynamics of organizational life.* Cambridge: MIT Press.
Hochschild, A. (1983). Comment on Kemper's "Social constructionist and positivist approaches to the sociology of emotions". *American Journal of Sociology, 89*(2), 432–434.
Hochschild, A. R. (1979). Emotion work, feeling rules and social-structure. *American Journal of Sociology, 85*(3), 551–575.
Hochschild, A. R. (2012). *The managed heart: Commercialization of human feeling.* Berkeley: University of California Press.

Hubbard, P. (2007). The geographies of "going out": Emotion and embodiment in the evening economy. In J. Davidson, L. Bondi, & M. Smith (Eds.), *Emotional geographies* (pp. 117–135). Farnham: Ashgate.

Huffington, C., & James, K. (2004). Containment of anxiety in organizational change: A case example of changing organizational boundaries. *Organisational and Social Dynamics, 4*(2), 212–233.

Hunter, B. (2001). Emotion work in midwifery: a review of current knowledge. *Journal of Advanced Nursing, 34*(4), 436–444.

James, K., & Huffington, C. (2004). Containment of anxiety in organizational change: A case example of changing organizational boundaries. *Organisational and Social Dynamics: An International Journal of Psychoanalytic, Systemic and Group Relations Perspectives, 4*(2), 212–233.

Kahn, W. A. (1992). To be fully there: Psychological presence at work. *Human Relations, 45*(4), 321–349.

Kendall, M., Murray, S. A., Carduff, E., Worth, A., Harris, F., Lloyd, A., et al. (2009). Use of multiperspective qualitative interviews to understand patients' and carers' beliefs, experiences, and needs. *BMJ: British Medical Journal, 339*, b4122.

Kenworthy, D., & Kirkham, M. (2011). *Midwives coping with loss and grief: Stillbirth, professional and personal losses*. Oxford: Radcliffe Publishing.

Klein, M. (1946). Notes on some schizoid mechanisms. In *Projective identification: The fate of a concept* (pp. 19–46).

Klein, M. (2011). *Envy and gratitude and other works 1946-1963*. New York: Random House.

Larson, E. B., & Yao, X. (2005). Clinical empathy as emotional labor in the patient-physician relationship. *JAMA: The Journal of the American Medical Association, 293*(9), 1100–1106.

Laurier, E., & Parr, H. (1999). Disability, geography and ethics. *Philosophy & Geography, 3*(1), 98–102. https://doi.org/10.1080/13668790008573700.

Lewis, P. (2005). Suppression or expression: An exploration of emotion management in a special care baby unit. *Work Employment and Society, 19*(3), 565–581.

Liaschenko, J., Peden-McAlpine, C., & Andrews, G. J. (2011). Institutional geographies in dying: Nurses' actions and observations on dying spaces inside and outside intensive care units. *Health & Place, 17*(3), 814–821.

Lief, H. I. (1963). Training for 'detached concern' in medical students. The psychological basis of medical practice. In H. I. Lief, V. F. Lief, & N. R. Lief (Eds.), *The psychological basis of medical practice*. New York: Harper.

Longhurst, R. (1997). (Dis)emboddied geographies. *Progress in Human Geography,**21***(4), 486–501.

Low, S. M. (2003). Embodied space (s) anthropological theories of body, space, and culture. *Space and Culture, 6*(1), 9–18.

Mackintosh, C. (2007). Protecting the self: A descriptive qualitative exploration of how registered nurses cope with working in surgical areas. *International Journal of Nursing Studies, 44*(6), 982–990.

Madden, R. (2017). *Being ethnographic: A guide to the theory and practice of ethnography*. Thousand Oaks: Sage.

Maddrell, A., & Sidaway, J. D. (2010). *Deathscapes: Spaces for death, dying, mourning and remembrance*. Farnham: Ashgate.

Mander, R. (1994). The care of the mother grieving a baby relinquished for adoption. Avebury.

Mann, S. (2005). A health-care model of emotional labour: An evaluation of the literature and development of a model. *Journal of Health Organization and Management, 19*(4/5), 304–317.

Mann, S., & Cowburn, J. (2005). Emotional labour and stress within mental health nursing. *Journal of Psychiatric and Mental Health Nursing, 12*(2), 154–162.

Mark, A. (2005). Organizing emotions in health care. *Journal of Health Organization and Management, 19*(4/5), 277–289.

Martin, P. Y. (2002). Sensations, bodies, and the 'spirit of a place': Aesthetics in residential organizations for the elderly. *Human Relations, 55*(7), 861–885.
Mattingly, C. (1998). *Healing dramas and clinical plots: The narrative structure of experience*. Cambridge: Cambridge University Press.
Mayall, M. (2010). A feeling for finance: Motivations for trading on the stock exchange. *Emotion, Space and Society, 3*(2), 103–110.
McCormack, D. P. (2005). Diagramming practice and performance. *Environment and Planning D: Society and Space, 23*, 119–147.
McCreight, B. S. (2005). Perinatal grief and emotional labour: A study of nurses' experiences in gynae wards. *International Journal of Nursing Studies, 42*(4), 439–448.
McDonald, S. (2005) "Studying actions in context: A qualitative shadowing method for organizational research." *Qualitative Research* 5(4): 455–47. https://doi.org/10.1177/1468794105056923.
Meier, D. E., Back, A. L., & Morrison, R. S. (2001). The inner life of physicians and care of the seriously ill. *JAMA, 286*(23), 3007–3014.
Menzies, I. E. P. (1970). *A functioning of social systems as a defense against anxiety*. London: The Tavistock Institute of Human Relations.
Milligan, C. (2005). From home to 'home': Situating emotions within the caregiving experience. *Environment and Planning A, 37*, 2105–2120.
Moss, P., & Dyck, I. (1996). Inquiry into environment and body: Women, work, and chronic illness. *Environment and Planning D: Society and Space, 14*(6), 737–753.
Moss, P., & Dyck, I. (1999). Journeying through M.E. Identity, the body and women with chronic illness. In E. K. Teather (Ed.), *Embodied geographies: Spaces, bodies and rites of passage*. Milton Park: Routledge.
Moylan, D., & Jureidini, J. (1994). Pain tolerable and intolerable: Consultations to two staff groups who work in the face of potentially fatal illness. In *The imaginative body: Psychodynamic therapy in health care* (pp. 222–240). London: Whurr Publishers Ltd..
Mumby, D., & Putnam, L. (1992). The politics of emotion: A feminist reading of bounded rationality. *The Academy of Management Review, 17*(3), 22.
Neyland, D. (2007). *Organizational ethnography*. Thousand Oaks: Sage.
Parr, H. (1998a). The politics of methodology in 'post-medical geography': Mental health research and the interview. *Health & Place, 4*(4), 341–353.
Parr, H. (1998b). Mental health, ethnography and the body. *Area, 30*(1), 28–37.
Parr, H., Philo, C., & Burns, N. (2005). 'Not a display of emotions': Emotional geographies in the Scottish Highlands. In J. Davidson, L. Bondi, & M. Smith (Eds.), *Emotional geographies* (pp. 87–107). Farnham: Ashgate.
Pile, S. (1991). Practising interpretative geography. *Transactions of the Institute of British Geographers, 16*, 458–469.
Pink, S. (2009). *Principles for sensory ethnography: Perception, place, knowing, memory and imagination*. Thousand Oaks: Sage.
Pink, S. (2015). *Doing sensory ethnography*. Thousand Oaks: Sage.
Porteous, J. D. (1985). Smellscape. *Progress in Physical Geography, 9*(3), 356–378.
Redinbaugh, E. M., Sullivan, A. M., Block, S. D., Gadmer, N. M., Lakoma, M., Mitchell, A. M., et al. (2003). Doctors' emotional reactions to recent death of a patient: Cross sectional study of hospital doctors. *BMJ: British Medical Journal, 327*(7408), 185.
Roter, D. L., Frankel, R. M., Hall, J. A., & Sluyter, D. (2006). The expression of emotion through nonverbal behavior in medical visits. *Journal of General Internal Medicine, 21*(1), 28–34.
Rowland, E. J. (2014). *Emotional geographies of care work in the NHS*. Unpublished doctoral thesis. London: King's College.
Rowles, G. D. (1978). Reflections on experiential fieldwork. In D. Ley & M. Samuels (Eds.), *Humanistic geography*. London: Croom Helm.
Simmel, G. (1997). *Simmel on culture: Selected writings*. Thousand Oaks: Sage.

Smith, P. (1992). *The emotional labour of nursing: Its impact on interpersonal relations, management and the educational environment in nursing*. New York: Macmillan.

Smyth, C. (2013, June 30). Nurses "too busy" to provide basic care for their patients. *The Times*. Retrieved from https://www.thetimes.co.uk/article/nurses-too-busy-to-provide-basic-care-for-their-patients-dq7r80s5hzv

Suchman, A., & Matthews, D. (1988). What makes the doctor-patient relationship therapeutic? Experiencing the connexional dimensions of medicine. *Ann Intern Med, 108*(1), 125–130.

Thompson, L. (2012). Deathscapes: Spaces for death, dying, mourning and remembrance. *Emotion Space and Society, 5*(2), 139–145.

Urry, J. (2000). *Sociology beyond societies: Mobilities for the twenty-first century*. London: Routledge.

Urry, J. (2005). The place of emotions within place. In J. Davidson, L. Bondi, & M. Smith (Eds.), *Emotional geographies* (pp. 77–86). Aldershot: Ashgate.

Widdowfield, R. (2000). The place of emotions in academic research. *Area, 32*(2), 199–208.

Willis, A. (2009). Restorying the self, restoring place: Healing through grief in everyday places. *Emotion, Space and Society, 2*(2), 86–91.

Wood, N., & Waite, L. (2011). Scales of belonging. *Emotion, Space and Society, 4*(4), 201–202.

Chapter 4
Case Study II: Care on the Move—The Emotional Geographies of Ambulance Crews

Emma Rowland

Abstract This chapter is a case study based on Emma Rowland's original empirical work that showcases some contemporary geographies in professional practice. It presents data derived from ethnographic observations of ambulance work and interviews with ambulance crew; the aim being to describe the nature and impact of particular spaces on this form of work including ambulance stations, mobile workspaces, public spaces and private spaces. Across these spaces, a number of experiences are mapped including, frustrations, stresses, coping strategies, gendered dimensions, and bodily agency. Indeed, this chapter uncovers issues related to professional health care work across non-traditional community spaces.

Introduction

Continuing the general aim of the last chapter—to showcase the geographies in professional practice—this chapter considers those in frontline emergency care (Rowland 2014). Indeed ambulance crews play an integral role in frontline emergency care, "assessing, managing, treating and transporting the public with an extensive range of conditions including falls, cardiac arrests, sudden death, severe trauma, [critical incidents] and social [or mental health] problems" (Minney and Bradley 2005; Blaber 2012; Williams 2013a, b, p. 208).

On the one hand, a significant factor and challenge is mobility. In caring for patients in the community, ambulance crews care work is spatially and temporally diverse as they move across the city. Previous research tells us that this mobility has a significant effect on crews' ability or desire to construct intimate emotional interactions with their patients, as well as affecting how they manage not only their own emotions at scene, but those of their patients, relatives and bystanders (Caroline 2007; Williams 2012a, b, 2013a, b, c). On the other hand, a significant factor is the situations encountered. Crews often attend life threatening, 'critical incidents' that

G. J. Andrews et al., *Place and Professional Practice*, Global Perspectives on Health Geography, https://doi.org/10.1007/978-3-030-64179-5_4

can evoke unusually strong emotional reactions in responders and possess the potential to interfere with their ability to cope at the scene or later in their lives (Halpern et al. 2009). Clohessy and Ehlers (1999) developed a list of critical incidents which evoked the greatest emotional reactions in ambulance crews. These included cot death, incidents involving children, dealing with the relatives of patients, burns and mental health patients and handling dead bodies. Alexander and Klein (2001) added to this list adding road traffic collisions, medical emergencies, suicides, violent incidents, the victim being known to the ambulance crew and the ambulance crew feeling helpless at the scene. Attending critical incidents impacts crew mates' constructions of intimate emotional interactions and enhances the emotional demands placed on ambulance crews.

Despite their important role in the health care system, ambulance crew are under-theorised and under-researched. As noted in Chap. 2, most geographical research has been on distributive features, routes and response times (Abbott 2008; Beillon et al. 2009; Mayer 1979; McMeekin et al. 2014; Ong et al. 2009; Peleg and Pliskin 2004; Uyeno and Seeberg 1984; Williams and Shavlik 1979). There is a diminutive body of research which focuses on the intimate social interactions and their emotion management skills in delivering care in the pre-hospital environment (Steen et al. 1997; Boyle 2005; Mitmansgruber et al. 2008; Filstad 2010; Brady and Haddow 2011; Brady 2012a, b, 2013, 2015; Clompus and Albarran 2016; Jennings 2017). It is this literature that the current study extends.

Despite ambulance crews being highly mobile, they have been neglected from the exploration of mobile working in the mobilities turn (Cohen 2010a, b). Instead, the mobilities turn (Urry 2003, p. 155) has focused largely on the role of mobile technologies and improvements to motorways and automobiles that have enabled workforces to become more mobile (Laurier 2004; Larsen et al. 2006). Examples include the car as a mobile office (Eost and Flyte 1998; Laurier 2002a, b), or on mobile occupations such as bus drivers and road inspectors (Normark and Esbjornsson 2005), cab and taxi drivers (Davis 1959; Psathas and Henslin 1967; Verrips and Meyer 2001; Mmadi 2013), lorry drivers (Agar 1986; Hollowell 1998), process engineers (Bertelsen and Bødker 2001), service technicians (Ueno and Kawatoko 2003; Wiberg 2005; Orr 2016) and mobile hairstylists (Cohen 2010a, b). Where the ambulance service have been studied, research tends to focus on the call handlers despatching the ambulance crews and the technology used to locate and despatch ambulance crews, rather than focussing on the crews working on the road (Ikeya 2003).

In these literary contexts, adopting an emotional geographies lens (as in the previous chapter), this chapter illuminates how the mobile nature of ambulance work impacts on intimate social and emotional interactions and emotional interactions with patients. It specifically examines how the spatial, temporal and mobile nature of this work impacts on crew's ability to manage, suppress, defend and cope with their own and their patients' emotions during the delivery of mobile patient care. Whilst this case study focuses on the care work of ambulance crews, it has translational application to other mobile and community-based care workers such as district/community nurses (Burke 2013).

Methods

Methodologically underpinned by (sensory) ethnography, this study sought to understand both the subjective and emotive experiences of ambulance crews as they deliver patient care on the move (Pink 2009, 2010, 2015). To understand the emic perspectives of the ambulance crew, it was imperative that I, the researcher (Emma Rowland), transitioned from an 'outsider' to an 'insider' (Headland et al. 1990; Parr 1998a, b; Herbert 2000; Van Maanen 2011; Bergman and Lindgren 2018). To support this process, bodily modification was initiated through the adornment of a fluorescent ambulance service 'observers' jacket. Whilst the jacket was supposed to help assimilate my body into the service, it simultaneously constructed my body to alternate from being both 'insider' and 'outsider'. For example, to patients and relatives, I was an insider, with some patients mistaking me for a paramedic in training. This mistaken identity was often beneficial as it enabled me to help crews at the scene by handing out equipment, fetching the stretcher or aiding with patient lifting and mobility. My identity was revealed once the patient was clinically stable to prevent any further assumptions. To the ambulance crews themselves, however, the observer's jacket was a clear distinguishing marker separating myself as a researcher from them in full ambulance uniform, and therefore my partially modified body initially hindered my transition to an insider. Initially, crews were wary of my presence, laboring under the misapprehension that I was a 'mole' observing their clinical practice. Over time, however, research relationships were established allowing integration into their organizational culture and the establishment of an insider perspective.

In collecting experiential data, ethnographic methods—shadowing (Bartkowiak and Sappey 2012)—and semi-structured and storytelling interviews (Gabriel 2000; King et al. 2018a, b) were implemented. Qualitative data were collected out of one ambulance station (Hermes) within one ambulance service in the South East of England Trust (SEAT). From this Trust, a range of ambulance crew personnel, including Emergency Care Assistants, Technicians, Paramedics, Emergency Care Practitioners, Operational Supervisors, Clinical Supervisors and the Station Manager, were accessed. Data were collected at the ambulance station, in the back of ambulances, out of Rapid Response Vehicles and public (parks, shopping centres, places of work) and private (homes, residential care homes) spaces.

Sixty-six crew mates were shadowed for 470 h, observing 203 patient journeys. Shadowing involved following ambulance crew mates for the duration of their shift. This was 12 h, if in a double manned ambulance, or 10 h, if single manned in a Rapid Response Vehicle. Crew mates were shadowed as they moved across the city (McDonald 2005; Neyland 2007), enabling a greater understanding of the crews' intimate social and emotional interactions with patients (Pink 2009). In the ambulance, ethnographic observations were taken from the 'observer's chair' which was situated at the head of the stretcher behind the driver's seat. From this vantage point, back and front region behaviours could be observed (Goffman 1980, 1990). When crews were not transporting a patient (at standby or en-route to a job for example), both crew members were sat in the front cab. From the observer's chair, conversations between crew mates could always be heard, and messages received from the call centre could be clearly seen on the multi-display terminal (MDT) situated on the ambulance's dashboard. When transporting a patient, the attending crew mate would be in the back of the ambulance, allowing mobile care work to be observed. At the scene (patient's homes, places of work, public spaces, etc.), observations were taken from a corner of the room or a space that was close enough to the action but not intrusive to the care that the crew needed to provide. Throughout all shadowing, ethnographic interviews took place to clarify crew behaviours, events and interactions. Further to the shadowings, static non-participant ethnographic observations (8 h) took place in the call centre (Davies 2008; Madden 2017). In total, two call handlers were observed. Semi-structured interviews and storytelling interviews (n = 4) were also conducted with Paramedics, Technicians and ECA's to gather a range of mobile care experiences and emotional narratives from crew mates at different operational levels. Interviews lasted on average 60 min, were audio-recorded and transcribed verbatim.

Ethnographic observations were recorded in a notebook (Atkinson et al. 2001; Hammersley 2007; Emerson et al. 2011). As mobile care work was unabating, notes were recorded contemporaneously. Due to the highly participatory nature of the research, on some occasions the notebook was an unwelcome, intrusive and obstructive tool and therefore was left in the ambulance. On these occasions, to ensure contemporaneous were still recorded, jottings were made on medical gloves. Ethnographic field-notes were typed electronically into ethnographic transcript over the next few days post observation. All qualitative data were inputted into ATLAS Ti 6.2 for data management (Friese 2019) and analysed using thematic analysis (Boyatzis 1998; Clarke et al. 2015). The following themes emerged from the analysis: the nature of mobile care work; emotional attachment in mobile carescapes; emotional detachment in mobile carescapes; and transforming people to patients. These themes structure the remainder of this chapter.

Mobile Care Work

As ambulance care work is spatially and temporally determined, it has an impact on the emotional care relationships which are established with patients on the road. Care work, which is performed in the patient's home, can result in ambulance crews establishing emotionally attached care relationships with their patients due to being surrounded by personal possessions, pictures and objects. In contrast, other spatialities such as public spaces encourage crews to become emotionally detached due to the lack of personalisation. Furthermore, due to crews' mobility, patient care is rarely performed in back regions (Goffman 1990). Instead, ambulance work is overwhelmingly performed in front of an audience (relatives, friends, colleagues or bystanders). This has implications for crew's emotional management (Goffman

1990; Hochschild 2012) as there is nowhere private to slip out of character or take a moment to breathe and refocus.

The technology used by crews is integral to their mobility. Throughout the shift ambulance crews use several technologies, both clinical and communicative, which enable them to work effectively in the pre-hospital environment. Communicative technologies include the multi-data terminal (MDT), which provides crews with information about the patient, taken from the call handler in the control centre. Information includes the patient's address, gender and age, presenting complaint and other relevant symptoms or illnesses raised by the person making the call. The MDT also allows crews to communicate with the control centre through the pressing of the 'to scene' button that shows the call centre that the crew has accepted the job. The satellite navigation (Sat Nav) is also linked to the MDT to direct the crew to scene. Crews may also use their in-car radio to request more information from control and a mobile phone, or to ask control for 'back-up' to convey the patient to hospital. In addition to communicative technologies, mobile clinical equipment is also integral to their mobile care work. In attending to a patient, crews take to scene a rucksack containing standard mobile diagnostic tests (blood pressure (BP), blood sugars, oxygen saturation, temperature), basic first-aid equipment (plasters, bandages, creams, medicines, etc.) and an oxygen tank which allows them to assemble a carescape upon arrival. Equipment attached to the ambulance may also be removed and brought to the patient's side such as suction and Automated External Defibrillator (AED) and equipment for mobilising patients such as collar and boards, scoops, stretchers and chairs. Innovations in technology have therefore enabled ambulance crews to treat patients more effectively and efficiently at scene or in the back of the ambulance en-route to hospital.

Spatialities of care work in the community are different from those in hospital care work. Emergency care work is unpredictable, unknown and often conducted in environments in which crews have limited control over due to being—"disorganised, chaotic and risky" (Watson et al. 2012, p. 648). These spatialities are often not designed for, and therefore ill-equipped to deal with medical situations. Arriving at scene, ambulance crews have to organise space by, for example, re-arranging or removing furniture to enable the delivery of care which involves complex "social, temporal and spatial management" (Watson et al. 2012, p. 648). The logistics of caring for a patient in the community can be challenging. Challenges are not only caused by the size of the space but also the temporality or urgency of removing the patient from the scene. Crews, therefore, have limited control over their carescapes, which may heighten crews' emotional labour, anxiety and stress.

Mobile care work is also isolating, and crews often lack the human resources to care for a patient in the pre-hospital environment. In the hospital environment, emergency care often requires several team members to manage the complexities of care. On the road, however, there are only two crew members in attendance, one if a Paramedic or Emergency Care Practitioner is single manned in a Rapid Response Vehicle. This means that crews must juggle logistical, clinical and emotional tasks simultaneously. This is in stark contrast to the hospital in which different health professionals can be called upon to treat various aspects of the patient. Dave Tolland an Emergency Care Assistant, commented:

> *It's entirely random, what we do, and it's uncontrolled... If you work in an Emergency Department (ED) your patients are in bed, they're lined out in front of you, all your equipment's nicely around you, there's plenty of staff to call on. Um, the patient's behaviour tends to be more controlled because they're in [a medical] environment. Obviously in someone's house or on the side of the motorway you've not got that much control. It takes a lot to get control. You really need quite a few people there, someone to control the relatives, someone to give you the equipment, someone to move furniture, or on the side of the motorway police officers to control the traffic. In hospital, you know everything is calm...if you need any drug, the pharmacist will come.... If you need an X-ray the radiographer will come... If you need their airway sorted, you just make a phone call and the anaesthetist comes. At the side of the roadway it's just you. If a relative visits a patient in hospital, there are a lot of sort of social cues that go with it. You go into the building, you know why you're there, it's a medical facility, all the doctors and nurses are wandering around in their uniforms, relative's in a bed, you know what you're getting. Whereas we deal with things that have gone suddenly wrong. The relatives aren't ready for it. They've just been sat at home on a normal day thinking nothing's going to happen, and then something goes wrong. We deal with a lot more raw feeling, especially from relatives, um, which can be very hard to deal with, because all your concentration is on the patient, um, and relatives who don't understand what's happened, aren't accepting it, don't know what's going on, become very alarmed all of a sudden...*

Emergency events which occur in the home are stressful and shocking for patients and their relatives as all are largely ill-prepared for such an event (Steen et al. 1997). Ambulance crews are therefore confronted with more raw emotions than hospital staff. Furthermore, there are specific social and cultural rules and expected behaviours which are associated with the hospital setting. These rules do not apply in the home, making treating patients and/or getting patients to comply with necessary treatment plans more demanding as crews must perform care within the patient's social and cultural contexts rather than abiding by the organization's rules.

As previously mentioned, the majority of ambulance care work is performed in front regions (Goffman 1990), with the audience typically comprised of patient's relatives or bystanders. In some circumstances the audience maybe participatory, helping the crew by providing information about what happened, the patient's medical history, medications and/or contact details for the next of kin. Providing care work in front of an audience is particularly challenging as there are limited spaces in which back region behaviours can be performed. The emotional labour required to sustain care performances in front regions is therefore exhausting.

During data collection, crews talked about the emotional pressure experienced in performing CPR in the home as relatives watch, scream and beg them to continue. In contrast, CPR in the hospital is usually performed in the resus room in the absence of an audience. In the event of unsuccessful resuscitation health professionals have time and (head) space to emotionally prepare themselves and consult additional members of the MDT before informing relatives. Ambulance crews do not have a back-region space in which to collect their thoughts and manage their emotions, nor do they receive training in the delivery of bad news (Steen et al. 1997, p. 60). This was emphasised by Dave Tolland, Emergency Care Assistant who commented:

> *Um, I've broken the news to families that their relative's deceased, which is a very hard thing to do... Compared to the hospital staff we have no training in it, and it's not like you can hide in resus for five minutes and prepare your speech... They've seen it in their own living room what you've done, and now you're telling them it's all over. I've had families and young kids screaming at us to keep going. You've got to break that and bring them down to earth. And you have to use quite horrible words. You can't say they've moved on. In denial, people won't accept that. You have to use words like dead, deceased. And you have to be quite firm, at which point you become the bad guy, because you become the focus for their anger, their rage. But you've got to do it*

Matthew Brand, paramedic, commented:

> *... I think the worst part about [resus] is the family around you and you know you have to tell them that their relative has just died and you know they are going to howl. That's the most difficult thing I think dealing with the howling, I feel sick, not physically sick like I'm going to vomit, but you know you get that sickness in the pit of your stomach, the gut wrenching knots in the stomach that you have to impart bad news and you know the reaction...*

Mobile care work demonstrates significant challenges for the performance of crews' emotions at scene. Due to the lack of back region spaces, crews are forced to sustain professional care performances in highly stressful and emotive situations. This has huge implications for their emotional wellbeing and ability to manage their emotions on the road.

Emotional Attachment and Mobile Carescapes

The majority of the literature on ambulance crews focuses on critical incidents (Steen et al. 1997; Filstad 2010; Williams 2012a, b, 2013a, b, c). Critical incidents, however, only take up 5% of ambulance crew jobs (Minney and Bradley 2005; Brady 2012a, b), with the majority of ambulance crews attending less traumatic jobs such as chest pain, falls or non-life-threatening injuries, which predominantly occur in the home. Delivering care in private spaces such as the home facilitates emotionally attached care behaviours as patients are surrounded by their family members and personal material objects, allowing the crews to see the individual person behind the incident. Some care work which is delivered in private spaces, is often less chaotic and less temporarily demanding, which further facilitates crews in establishing intimate and emotional care relationships with their patients and their relatives.

In response to the absence of research focusing on the less critical/serious jobs attended by crews, the following ethnographic extract with paramedic Matthew Brand and Emergency Care Assistant Jeff Osborne provides insight to the routine and everyday care work provided by crews. It demonstrates that due to a lack in temporal restraints (non-emergency care) ambulance crews are able to establish intimate emotional care relationships with their patients (Steen et al. 1997; Filstad

2010; Williams 2012a, b). In the following extract, the patient, Gordon, was treated for a suspected fractured neck of femur (NOF)—a broken hip after tripping over his garden hose. This extract—of shadowing of Matthew Brand, Paramedic—emphasises how the spatialities in which care is provided causes not only logistical obstacles but also impacts on the establishment of emotional care relationships with patients.

The MDT bleeps 'M 81 fallen, hip pain NOF? Age 81'. Jeff comments 'bless him' ...The crew arrive at scene and are led through the house to the back garden by the patient's wife... In the garden the patient is sat on a kitchen chair wrapped in a blanket. His daughter is crouched by his side holding his hand and rubbing his shoulders in what seems an attempt to warm him up. Jeff greets the patient warmly and crouches down in front of him as he introduces himself... Jeff asks, 'what happened this evening then?' Gordon says that he is a 'silly old fool' who fell over whilst watering his plants. Jeff establishes whether Gordon had felt dizzy or tripped.

Gordon explains that the hose pipe got stuck on something and that he had tugged hard on the pipe and it had freed itself quickly causing him to lose his balance... and tumble backwards. He says that he has pain in his leg and wrist. Gordon comments that his wife (Sylvia) is always complaining that he is too old to be out doing the entire garden and this fall has proven her right. He whispers to Jeff 'don't tell her I said that'. Jeff smiles at the patient and promises that he won't. Jeff attaches a sats probe to the patient's finger and wraps a BP cuff around his right arm...

Jeff asks Gordon if he can look at his leg. Gordon nods and Jeff un-wraps the blanket from around his body and legs leaving it hanging from his shoulders. Jeff comments to Matthew that it looks rotated... and the wrist is very swollen...Sylvia exits the house and stands next to Matthew and provides him with Gordon's previous medical history... Matthew asks Gordon if he would like some pain relief. He would, as he has 10/10 pain...

Matthew goes out to the ambulance to get the morphine asking Jeff to prepare a flush... Jeff opens the rucksack on the grass and begins preparing the needles and syringes and flush and lines them up on the path in front of Gordon. As he does so he checks the dates on the bottles and talks informally to the patient...

Matthew returns with the morphine... preparing the canular, he talks to the patient about his vegetables, commenting on good crops and flourishing plants. He asks what his secret is as he has failed miserably to grow anything in his own garden...Matthew checks the date on the morphine and then...as he administers one shot comments 'great shed'. Gordon tells Matthew that every man needs a good shed. Matthew asks what he does in there as he writes the dose and time on his glove (20:46 – 2.5mg)...Gordon says he mostly does potting but also, he makes wines. Gordon asks Sylvia to get the crew a bottle each to take with them but Matthew declines). Matthew asks Gordon what he had been in hospital for and Gordon says that he had cancer... in a break in the conversation Jeff asks Gordon what his pain score is now. Gordon says that it has gone down to 4/10. Matthew slowly administers another shot (5mg) at 20:53...

With the pain reducing Matthew and Jeff discuss how they will get Gordon onto a stretcher. The patient is sat on his chair on a narrow path surrounded by a vegetable bed and flower beds. It is a logistical challenge and Matthew returns to the ambulance to retrieve a rotunda (equipment that enables them to pivot patients on the spot). Matthew administers one more shot of morphine and then he and Jeff stand either side of Gordon with their arms around his waist and his arms around their shoulders...on Jeff's count of three they help the patient to his feet...the patient is swivelled 180 degrees so that he has his back to the stretcher...Matthew and Jeff lift the patient up onto it, Gordon's body pressed against theirs for support...Gordon winces in pain and lets out a small moan.

On the stretcher Jeff and Matthew quickly work around Gordon to make him comfortable...the crew can now see that Gordon's left leg is twisted and extended. Jeff informs

> *Gordon that he does have a broken hip... Matthew asks the patient whether he thinks that he needs any more morphine. The patient says that he is now comfortable and that the pain score is now 1/10...Gordon is wheeled towards the ambulance.*

This extract raises two discussion points covered next: the role of labelling and categorising patients and the need to establish rapport required to facilitate intimate and emotionally attached relationships.

Regarding the labelling and categorisation, as the extract opens, the ambulance crew made judgements about this patient based upon the information which came through the MDT. Jeff labels Gordon a "good patient" deserving of their care, emphasised by his comment "bless him" (Stockwell and Stockwell 1972; Kelly and May 1982; Maben et al. 2012). In other observations, crews were heard to judge jobs as "a load of rubbish" or "a waste of time". A patient's illness or medical condition, age or social class can affect health professionals' attitudes towards their patients (Kelly and May 1982) with instant judgements tending to affect how crews treat patients at scene (Williams 2013a, b, c).

Labelling patients as 'good patients' (who were deserving of their care) helps crews to start building a positive emotional relationship with their patients. Establishing rapport and an emotional care relationship with patients is an important aspect of crew work. However, due to the focus upon critical incidents or 'bad patients', which require emotionally detached behaviours, emotionally connected care work has been sorely neglected by the literature (Steen et al. 1997; Brady 2012a, b; Williams 2012a, b, 2013a, b, c). In building emotional relationships with their patients in the community, ambulance crews must view the patient in front of them as a person rather than just an ailment requiring fixing. Therefore, they need to be skilled social actors, able to adapt to different social classes, religions and cultures and to be able to put their patients at ease as soon as they meet them. This helps crews to build trust, empathy and an emotional understanding of the situation, which is often in stark contrast to care relationships in the hospital. This was emphasised by Paramedic Cheryl Hooper:

> *We're better at dealing with the person and the family than hospital staff. I think it is because they don't see them in their own environment. I think one of the hardest things is when you get to the ED and their [symptoms] are brushed off as nothing. For that person and that family, [being admitted to hospital] is a massive significant thing...I just want to say to them it maybe another CVA (Cerebral Vascular Accident) to you...but this chap was normal yesterday, really independent...this is really going to affect him and his family... they just don't see the bigger picture...it's very frustrating.*

Cheryl emphasises that the spaces in which ambulance crews deliver care facilitates emotional attachment, and, in contrast, because ED staff do not see patients in their home environments, they don't see the "bigger picture". Arriving at the ED patient's social and familial circumstances are often deemed irrelevant to ED staff in treating the presenting complaint. This information, however, becomes integral to the patient's discharge but is often unrecorded and therefore lost. This is a source of great frustration to ambulance crews because their time and care invested in the patient and their relatives is perceived to be taken for granted and devalued.

Returning to the extract with Gordon, Matthew and Jeff, we see that building rapport with patients is integral to mobile working, especially in jobs which are not time-critical or temporally demanding where ambulance crews may be in the patient's home for hours. To keep the interaction with their patients personal, ambulance crews write the name of their patient on their gloves. This was emphasised in an ethnographic interview with student paramedic Jason Brand who stated "the first thing that goes on the glove is the patient's name, you will see I have to keep looking at the glove to remember their name, you know, for the personal touch".

As Matthew and Jeff treat Gordon, they extend their relationship by asking him personal questions about his garden. These questions not only fill the time and make the care-interaction less awkward, they also help ambulance crews to build an understanding of the patient's social circumstances, which may be important to their care, enabling them to treat the patient more effectively at scene but also to provide the ED with more holistic information.

Additionally, establishing relationships with patients' relatives is also important for crews as they can provide vital information which enables the crew to treat the patient more effectively. In the extract, for example, Sylvia provides crew with details about Gordon's previous medical history. Whilst the relationship with the relatives is important, current literature fails to report the benefits of relatives at scene. Instead, it focuses on the challenges of relatives at scene, especially during critical incidents such as cardiac arrests (Brady 2012a, b; Williams 2012a, b, 2013a, b, c). Relationships with relatives, however, can reduce the emotional burden, or anxieties crews face by providing highly relevant medical information which supports their diagnosis and treatment.

Emotional Detachment in Mobile Carescapes

In contrast to the care work performed in private spaces, such as the home, care work performed in public spaces - such as shopping centres, bars and restaurants, hotels, high streets and recreational spaces - encourage emotionally detached care relationships between crews and their patients. In public spaces crews do not become immersed in the patient's social and emotional worlds, because they are not surrounded by their personal material objects and possessions which allows them to quickly establish rapport with their patients. Furthermore, the types of jobs attended to in public spaces also enhance emotionally detached care behaviours as critical incidents (road traffic collisions, suicides, medical emergencies and dealings with drunks and drug abusers) are more likely to occur. These types of jobs are more likely to require crews to engage in task orientated behaviours (Menzies 1970).

In the following extract with Paramedic Bobby King and Technician Josh Ledger, we witness the emotional labour required to provide patient care to those patients who have been labelled deviant, time wasters or 'bad patients'—for example, drunks, drug abusers, abusive patients. In this extract, Bobby and Josh spend 3½ h with a drunken verbally and physically abusive teenage girl:

We arrive at the edge of parkland at 00.11…a Rapid Response Vehicle is already parked-up and an Emergency Care Practitioner, Steve Blyth, is leaning against his car bonnet, his arms folded across his chest looking unimpressed. Two girls are lying on the floor huddled over each other covered in blankets… Steve walks towards the ambulance and hands over a Patient Report Form (PRF) and… says: 'Fairly pissed, not able to walk. Basically, they both just need to go home. I have lost interest (patient vomits loudly on the ground) Oh magic, they are being sick again!'…

We walk over to the girls…The blankets are barely covering them, their dresses ridden up to their waists. The larger girl, Kate has huge holes in her tights with a bum cheek protruding. Bobby laughs and shakes his head 'what a view!'

Josh crouches down in front of the girls and tries to get their attention…and suggests that they get into the ambulance. Kate tries to drunkenly encourage Zoe (patient) to open her eyes and stand up. There is a lot of giggling and swearing as the patient refuses to get up. Zoe begins retching and Kate pushes her head away and she is sick on the floor. Zoe wipes her mouth with her hand spreading vomit over her face and hair. Josh declares loudly and firmly 'time to get up!' He takes the patient's arm and pulls her to a standing position. Bobby helps Kate to her feet…Zoe swears abusively 'get off me you cunt…you're a twat leave me alone'. Kate laughs and tells her not to be rude but her laughing only encourages Zoe's abusive language. Bobby firmly tells her to stop with the language as he is just trying to help her. In the ambulance…the girls are wrapped in clean blankets the vomit covered blankets are left in a pile on the floor…The patient begins heaving again and Josh quickly gets a vomit bowl from the cupboard above her head and thrusts it in front of her face. She pushes it away but Josh thrusts it under her chin again. Zoe pushes it away again. Bobby takes the bowl from Josh and holds it firmly under Zoe's chin 'you won't be sick on my ambulance floor…or you can clear it up!' The patient laughs 'fuck off you twat!' Kate tells Zoe off 'I'll hold the bowl'. The patient takes the bowl and is sick. Bobby hands Kate tissues and she wipes Zoe's face.

Kate asks the crew if they can drive the patient back to her house where she can look after her… Zoe tries to get up. Josh puts his hand firmly on Zoe's shoulder and tells her to sit. Josh asks Zoe where she lives. She gives some vague description of a place opposite a Church. He presses her for a road name. After a lot of giggling and swearing she finally comes out with a name of a road and Bobby looks it up in the AtoZ whilst Josh probes her for a house number…

Bobby takes Zoe's Blood Pressure (BP), BM and attaches the sats probe to her finger which she keeps ripping off and Kate keeps re-attaching… The patient moans that the BP cuff is hurting her and tries to rip it off. Unable to she gives up and slumps over with her head on her friend's lap…Bobby climbs into the front of the ambulance and we leave the scene. Josh and I sit on the stretcher…throughout the journey and Josh tries to coax a story from Kate about how much they had to drink… Zoe talks incoherently and giggles. Josh looks at me and rolls his eyes. Kate then begins to blow raspberries on Zoe's face, and they laugh. Josh and I look at each other, eyebrows raised… Nearing the housing estate…Bobby asks Kate to direct him…Arriving at the flat Bobby opens the back of the ambulance. Kate tries to help Zoe stand but she keeps flopping over and sitting back on the chair with her head slumped onto her chest. Josh asks Kate to sign the PRF to state that the patient has been left in the care of a friend. With the PRF signed Josh hands me the form and asks me to put it in Zoe's handbag so it doesn't get lost….

Bobby and Josh help the girls out of the ambulance. Zoe refuses to walk and lies down in the road. Bobby tells her firmly to 'stand up' but she shouts abuse at him. Kate asks her where her key is, and she says it is in her purse. I look through the purse but there is no key. Kate snatches it from me and tries to find it herself…The key can't be located. A neighbour comes out of his house and says that he can let them into the main building. The crew thank him. Bobby asks the patient again to stand but she refuses so lifts her up from under her armpits. She continues to shout abuse at him kicking violently. Bobby hauls her to her feet…

she lifts her legs like a child that refuses to walk. Bobby tells her abruptly 'walk'. Her verbal abuse towards Bobby continues and he sighs and carries her to her front door. She crawls inside...

The crew return to the ambulance. They sit and watch what is going on inside the flat. Kate has disappeared leaving Zoe lying on the entrance mat. They discuss what they should do. Bobby states 'we can't leave her there... Tomorrow we will see in the [paper] Ambulance crew leave girl in hallway to die'. They suggest they sit and wait for a couple of minutes to see if her friend comes to get her. After a short while Kate and the neighbour enter the entrance hall and kneel down in front of Zoe. They are there for a couple of minutes before the neighbour comes out and waves us over...The neighbour says that the patient is fitting.

The crew enter the flat, the patient is shaking...There is a wet patch on the carpet where she has wet herself. Bobby asks the friend about the 'fitting'. Kate says that her friend has no history of epilepsy. Bobby tells Kate that he does not think that she has had a fit and she is more likely to be shaking violently because she is not wearing any clothes and it is a cold December night. As Zoe can't get into the flat...she will have to go to hospital...

The patient swears at Bobby saying she doesn't want to move...Bobby carries her like a child to the back of the ambulance but she refuses to get in, hitting and kicking Bobby in the vain hope that he will put her down...Josh climbs into the ambulance and tries to coax her inside. Bobby lifts her and tries to hand her to Josh but she kicks out her legs and places them on either side of the door shouting more abuse. Bobby asks her to stop shouting as it is waking up the neighbours, she says that she doesn't care and continues shouting. Eventually she is in the back of the ambulance and they strap her into the stretcher her arms and legs flaying violently lashing out at Bobby...

Zoe shouts for Kate...The crew make the decision to leave without her as they think that it will be worse in the ED if they are together. Bobby leaves the scene and we head towards hospital at normal road speed...The patient lays on the stretcher face down with her head leaning over the side pretending to retch and occasionally shouting obscenities...

(03.42) We arrive at the hospital and Bobby opens the back door of the ambulance. He tries to take off the stretcher straps to allow her to walk into the ED, but she screams 'you're a fucking cunt, you fat twat!' and kicks her leg out towards him where it makes violent contact with his stomach. Bobby retorts angrily holding his stomach 'and you're a silly little drunk girl!' Instantly Bobby moves back from the patient and briefly closes his eyes and exhales. He apologises for shouting at her. She tells him she will make a complaint and get him fired and continues to use abusive terms towards him trying to get another reaction. Bobby suggests they just wheel her out on the stretcher....

We return to the ambulance where Bobby apologises... and says that he should not have reacted. He is extremely frustrated with himself.

In this data, 'establishing emotional detachment' comes to the fore. In treating 'unpopular' and 'deviant' patients (Stockwell and Stockwell 1972; Kelly and May 1982; Maben et al. 2012), crews adopt coping strategies such as "self-control, distancing and avoidance" (Thompson 1991; Clohessy and Ehlers 1999) to become emotionally disengaged with their care work. In response to Zoe's antagonistic and verbally aggressive behaviours, Bobby and Josh carefully manage their emotions and emotionally detach themselves in two ways. First, they create physical distance from Zoe by limiting their bodily proximity; and second, personal engagement, or 'verbal proximity', is limited (Malone 2003). Despite careful emotion management, however, Zoe's unrelenting deviant behaviour "takes an emotional toil" on Bobby, resulting in an emotional rupture (Mitmansgruber et al. 2008).

The creation of physical distance is observed by all ambulance crew attending Zoe and Kate. As the extract opens, we witness Emergency Care Practitioner Steve Blyth waiting at a distance from the girls. This distance signifies Steve's emotional detachment from the patient and his bodily deportment, arms crossed against his chest, which emphasises his lack of interest, anger and frustration. Kelly and May (1982) found that drunken or alcoholic patients evoke negative responses and emotions in crews because they prevent them from using their clinical skills. Instead, the perception is that crews are used as a glorified taxi service to transport patients to the ED, and this takes them off the road so that they are unavailable to attend deserving patients (Papper 1970; Jeffery 1979; Kelly and May 1982) as we witness in this extract. The lack of bodily proximity is continued when Bobby and Josh take responsibility for these patients with both crew members pulling the girls up off the floor by their arms, thrusting a vomit bowl under Zoe's chin asking Kate to clean up Zoe's face. The way in which Bobby carries Zoe also increases bodily distance. This bodily disconnect is observed in stark contrast to the close bodily proximity observed in the mobilisation of Gordon.

In addition, the crew establish emotional detachment though their lack of personal engagement with Zoe and Kate. Unlike Matthew and Jeff, Bobby and Josh do not engage in any personal conversations with their patients. Instead, they only converse with the patients to establish what and how much the girls have had to drink and the directions to their house. Furthermore, the crew's tone of voice is often assertive and strong with a 'sombre demeanour' (Boyle 2005) to demonstrate that they do not think that this behavior is appropriate and should be tolerated. Ambulance crews also often use humour to cajole 'difficult', intoxicated or 'irate' patients into compliance (Boyle 2005). Humour is also a safe way to vent anger directed at patients and act as a defence mechanism against frustrations, enabling crews to gain a sense of power and superiority over poorly behaved patients through the 'role distancing technique' (Goffman 1990; Marra and Holmes 2002; Tracy et al. 2006). We see in this extract Bobby using humour to relieve his frustration as he laughs at the girls' lack of dignity—*'what a view!'*. Humour and jokes can reduce the possibility of an emotional rupture. On this occasion, however, humour is not enough with the interaction with Zoe. Bobby exposes his authentic emotions through an emotional rupture (Goffman 1990; Hochschild 2012). This emotional slippage is unsurprising considering the length of time that the crew had endured this patient's antagonistic behaviour, abusive language and physical violence within the small confines of the ambulance. The longevity of the crew's tolerance demonstrates the difficult emotional labour required to perform and manage emotions through surface acted behaviours. It also emphasises how surface acted performances can easily slip causing a professional faux pas.

Immediately following his emotional rupture, Bobby takes a step backwards to increase his physical distance from the patient and takes a moment to regain his composure and collect his thoughts so that he can recover his emotions. Bobby is however incredibly frustrated and angry with himself for not being able to sustain his emotionally detached professional performance and for allowing Zoe to coax his authentic emotions out of him.

Whilst the extract demonstrated how crews create emotional detachment with deviant patients, interview extracts with ambulance crews illustrated the importance of emotional detachment in life threatening situations (Boyle 2005; Minney and Bradley 2005). Emergency Care Technician Dave Tolland, for example, explains how crews emotionally detach themselves from their patients, to deliver "brutal interventions such as ramming tubes down patient's throats, cracking ribs…drilling into bone":

In those few cases where we've actually really got to intervene… I don't see them as a person. I just see them as various systems that I can fix…We went to a gentleman yesterday who, um, was on his last few minutes...his airway was completely obstructed with vomit, and it was like right!... so you just immediately go straight into your system...we just pushed him over, slapped his back, a colleague ran to get the suction, got an airway in, suctioned out loads [of vomit], rolled him on his side, put the oxygen on, and you're just working through...the checklist in your head...[you need to] stabilise, monitor, then get them out of the house… it wasn't until he was on the resus bed …when we stop and go, ooh, that was a really good job… But when you're on the job you're just thinking what's next? And that's it…focussed on the task…

Dave reveals that, for those 5% of life threatening calls made to the ambulance service (Minney and Bradley 2005; Brady 2012a, b), ambulance crews emotionally detach themselves from the patient by viewing them as a set of body parts that requiring fixing (Palmer 1983; Regehr et al. 2002; Scott 2013). This was also discussed by paramedic Matthew Brand who likened delivering medical interventions to a mechanic fixing a vehicle.

When I've got a poorly patient that's not quite dead, or might be on their way there, then I'm thinking…no different to how I'd look at an engine…I'm thinking about the treatment… how can I fix this?...once I stop treatment, I try and talk to them and let them become a person again, you know, but for…a little while they are something that just needs fixing and we're trained in that way almost...like cars and motorbikes...an engine...erm, a heart...if that's not working we need to fix it

Matthew and Dave's experiences of coping with critical incidents demonstrates that ambulance crews get "tunnel vision" (Boyle 2005, p. 64), which allows them to focus on the patient as an object (Filstad 2010; Williams 2012a, b, 2013a, b, c). This enables crews to deliver the medical interventions required rather than on the brutality or the pain and injuries that they may cause the patient. Dave also talks about following processes and checklists in his head that allow him to focus his mind on the task in hand rather than thinking about the emotionality of the incident. Williams (2013b) revealed that paramedics "slavish adherence to established protocols and procedures help them to emotionally detach themselves from the patient" (p. 516). Indeed, following Joint Royal Colleges Ambulance Liaison Committee (JRCALC) guidelines serves as a defence against the anxiety and guilt of treating patients in mobile carescapes (Scott 2007).

Observations and interviews have illuminated the emotional detachment initiated in mobile workplaces. Ambulance crews, however, were also observed to instigate emotional detachment with those patients who they had initially created emotionally attached care relationships with at scene as they conveyed them to hospital.

Creating emotional distance en-route to the hospital was often an unconscious process but was necessary for establishing emotional closure.

Transforming People to Patients

In the following extract, we return to ethnographic observation with Gordon, Matthew and Jeff to demonstrate how ambulance crews use their mobile workplace to begin the transition from an emotionally attached care relationship to an emotionally detached care relationship as they transport patients to hospital. We witness how the ambulance en-route to hospital becomes an invaluable space for disconnecting emotional relationships, allowing crews to successfully move on to the next patient. Returning to the extract we witness Gordon being wheeled into the ambulance:

> *At 21.20, Gordon is wheeled onto the ambulance....Jeff wraps the ambulance BP cuff around his arm and clips the sats probe to his finger. Jeff looks at the mobimed and says aloud that Gordon's sats are 87% and his pulse is 86. Jeff gets an O2 mask out of the drawer and attaches it to the ambulance's oxygen supply. He puts the mask over Gordon's face. Matthew sits down and begins filling in the PRF...*
>
> *...Matthew hands Jeff the PRF and asks him if he is happy to go. Matthew climbs out the back of the ambulance and slams the back door. Jeff writes the time directly onto his PRF and then asks Gordon for his surname, address, home contact number, details of his next of Kin, GP, GP surgery...Jeff then talks through the patient's previous medical history with him...*
>
> *...Jeff looks up at the mobimed and states that his sats are now up to 96% so removes the oxygen mask. Jeff continues to fill in the PRF... frequently asking medical questions.*
>
> *PF You say your shoulder was giving you some pain?*
>
> *P Not much really, a little*
>
> *PF Well we will mention it, anyway, get it looked at. And all your injuries are down your left-hand side*
>
> *P Yes, that's right*
>
> *PF So it looks like you took all your weight on that side...you haven't got any injuries elsewhere?*
>
> *P No, I kind of rolled when I went down but only have pain on the left*
>
> *As Jeff writes in the text box, he copies some information across from his glove. He then looks up at the mobimed and says that his oxygen levels have dropped again. He asks, 'would you like a bit more oxygen?'... Jeff continues with the questions:*
>
> *PF Any allergies?*
>
> *P No....*
>
> *PF Have you taken any paracetamol today?*
>
> *P Yes.*
>
> *Jeff examines Gordon's wrist and fingers and comments on the swelling...Gordon tells Jeff a story of how he broke his wrist the second time (trying to help his daughter pump a flat bike tyre when he fell). Jeff nods and responds occasionally with grunts. He then leans over Gordon to take Gordon's BP once more (168/72). Jeff asks whether Gordon hit his head and or had been knocked out. Gordon replies no and Jeff continues to write in the text box. Gordon comments 'I can't get over how stupid I am...all the family will say 'silly old fool shouldn't have been doing it!' But I enjoy the garden you see...' Jeff doesn't respond, stands and asks Gordon if he can have a feel of his shoulder commenting 'we don't want to*

miss anything do we?' Jeff examines the shoulder and then takes a seat and continues writing.

We arrive at hospital at 21.48... the hospital is very busy with a queue of 6 stretchers lined up along the corridor. We join the back of the queue. Jeff apologises to Gordon and says that it looks like it is going to be a long wait.

...At the front of the queue Jeff hands over Gordon to Natalie Taylor (NT) the Rapid Assessment Triage (RAT) nurse

NT Name?

PF Gordon Philips

NT (spells aloud as he types) DOB?....ok what's wrong?

PF Our patient was watering his garden with a hose pipe when he fell. Query NOF and fractured wrist

NT Is it short and rotated?

PF Yes shortened and rotated at ankle

NT Does he have a pulse there?

PF Good pulse, suffers from circulation problems. Previous medical history - prostate cancer, arthritis.....

NT You can go into 1...just need to move that patient out...can you wait a moment?...

Patient in bay 1 is removed by two nurses and Gordon is wheeled into the bay. The ward clerk enters the bay and takes the PRF from Jeff and walks over to her desk. A PAT slide (patient transfer slide) is used to move Gordon from the stretcher to the hospital bed. Matthew wheels the stretcher out of the bay and removes the linen and throws it in the sluice. He wipes it down with some anti-bacterial wipes and heads to the linen cupboard and places fresh linen on the stretcher... with the stretcher ready for the next patient we leave the hospital.

Earlier we heard how Paramedic Cheryl Hooper expressed her frustrations with the ED staff for not being able to see the bigger picture surrounding presenting patients and their failure to see their patients as people. The extract, however, as suggested shows how ambulance crews' emotionally attached care behaviours diminish in the ambulance as they mobilise a patient to hospital. In contrast to the care behaviours at scene, Matthew and Jeff begin to transform Gordon from a person into a patient. While for most ambulance crews this process of depersonalisation was unconscious, Emergency Care Assistant Dave Tolland was conscious of how his behaviour and clinical procedures inside the ambulance resulted in the depersonalised patient.

We see a lot more personal identity [than ED staff], we see their home and we see their pictures, we see their family, we see their pets. Um, we see how they live, we see them I guess at their worst...but by the time they get to hospital we've cleaned them up, wrapped them up [in blankets]...sorted them...packaged them, made them look like a patient...put bandages on them and a set of numbers on them...so yeah, you know, hospitals get presented with patients, we take them from being [John Smith] from [6 Hassock close] and turn them into a patient, a [NHS] number.

In alignment with Dave's explanation, Jeff transforms Gordon into a patient. He is wrapped in a blanket and has ambulance diagnostic equipment attached to his body. Through the placement and insertion of medical technologies, Gordon's body is turned into a patient cyborg (Haraway 1994; Lapum et al. 2012a, b). In addition, Jeff strips Gordon of his name by completing the PRF and turning him into an incident number. Furthermore, along the journey we observe how Jeff's interaction begins to change. Jeff becomes more concerned with finding out Gordon's medical history than in engaging with him on a personal level. On two occasions Jeff ignores Gordon's initiations of personal anecdotes turning his attention to Gordon's blood pressure reading. This disengagement illustrates the process of emotional detachment. Emotional detachment culminates in the hospital when Jeff no longer calls Gordon by his name, other than to book him in. Instead Jeff refers to him as "our patient". Matthew and Jeff's emotional detachment is heightened as Gordon is transferred from the ambulance stretcher to the hospital trolley. Gordon is then symbolically erased from Matthew and Jeff's memories through the removal of the stretcher's linen and the wiping down of the stretcher with anti-bacterial wipes, before new linen is placed on it. Finally, the crew discard their gloves containing his name, date of birth and other personal information thus removing all traces of him. The significance of removing gloves was emphasised by student paramedic Jason Bond:

> *The first thing that goes on the glove is the patient's name...the minute the glove is removed the name is forgotten...by the end of the day we might have seen 5-8 patients...I doubt anyone can remember the name of their last patient, let alone the first.*

The transitioning of named people to patients demonstrates how integral the mobile workplace is to crews' emotion management and facilitating emotionally detached care behaviours. The spatiality and the temporality of the ambulance assists crews in emotionally disconnecting from their patients through the application of mobile technologies and incident numbers as they mobilise from scene to the hospital. The process of emotional detachment serves to reduce the emotional labour required by crews, enabling them to reduce their emotional management and engage in back region behaviours before they are redeployed to the next incident.

Conclusion

This chapter explored care mobilities, or care on the move, specifically articulating the intimate interactions of ambulance crews and related emotional attachments, detachments and patient transformations. The data exposed microsocial conditions of care highlighting, for example, that bodily proximity allows crews to build rapport or 'verbal proximity' with their patients, which meant that they were able to get to know their patients more intimately, helping them to treat and manage their patients more rapidly at scene as well as aiding their clinical decision making. More generally, whilst care work undertaken in private spaces such as the home was observed to induce intimate interactions and emotionally attached care relationships (due to crews being surrounded by patient's families material objects), critical incidents attended in public space were observed to encourage emotionally detached care relationships. Temporality also affected the establishment of emotional care relationships with the day of the week and time of day influencing the types of jobs and patients attended and affected intimate interactions with patients. As suggested, these geographies are important in themselves, but might also resonate with aspects of other forms of community-based professional practice.

This research is the first to navigate and explore the importance of space, place and proximity to care delivery and consumption of mobile care work. It therefore combines narrative from workplace mobile geographies to analyse routine mobile care work and its emotional management by health care workers. Whilst much knowledge is well-established in geographical health professional research, the care work of mobile health professionals, particularly ambulance crews, is in its infancy and is yet to find a home within geographical inquiry. Further investigation is required to elucidate the spatial and temporal dynamics of place in constructing and de-constructing emotional care relationships between health professionals and patients and the impact of place on the emotional labour of these health professionals.

References

Abbott, R. M. (2008). Geographic analysis of ambulance availability in Arkansas. *Family & Community Health, 31*(1), 35–43.

Agar, M. (1986). *Independents declared: The dilemmas of independent trucking* (Smithsonian series in ethnographic inquiry). Washington: Simthonian Institution Press.

Alexander, D. A., & Klein, S. (2001). Ambulance personnel and critical incidents: impact of accident and emergency work on mental health and emotional well-being. *The British Journal of Psychiatry, 178*(1), 76–81.

Atkinson, P., Coffey, A., Delamont, S., Lofland, J., & Lofland, L. (Eds.). (2001). *Handbook of ethnography*. Thousand Oaks: Sage.

Bartkowiak, I., & Sappey, J. (2012). The methodological identity of shadowing in social science research. *Qualitative Research Journal, 12*(1), 7–16.

Beillon, L. M., Suserud, B. O., Karlberg, I., & Herlitz, J. (2009). Does ambulance use differ between geographic areas? A survey of ambulance use in sparsely and densely populated areas. *The American Journal of Emergency Medicine, 27*(2), 202–211.

Bergman, Å., & Lindgren, M. (2018). Navigating between an emic and an etic approach in ethnographic research. Crucial aspects and strategies when communicating critical results to participants. *Ethnography and Education, 13*(4), 477–489.

Bertelsen, O. W., & Bødker, S. (2001). *Cooperation in massively distributed information spaces*. New York: Springer.

Blaber, A. (2012). *Foundations for paramedic practice: A theoretical perspective*. New York: McGraw-Hill Education.

Boyatzis, R. E. (1998). *Transforming qualitative information: Thematic analysis and code development*. Thousand Oaks: Sage.

Boyle, M. V. (2005). “You wait until you get home”: Emotional regions, emotional process work, and the role of onstage and offstage support. In C. Hartel, N. Ashkanasy, & W. Zerbe (Eds.), *Emotions in organizational behavior* (pp. 64–84). Hove: Psychology Press.

Brady, M. (2012a). Pre-hospital psychosocial care: Changing attitudes. *Journal of Paramedic Practice, 4*(9), 516–525.

Brady, M. (2012b). Using patient experiences of emergency and unscheduled care for quality improvement. *Journal of Paramedic Practice, 4*(10), 576–584.

Brady, M. (2013). Health and care professions council: Protecting whom. *Journal of Paramedic Practice, 5*(5), 246–247.

Brady, M. (2015). Death anxiety among emergency care workers. *Emergency Nurse, 23*(4), 5–16.
Brady, M., & Haddow, P. (2011). Evolving education in paramedic practice. *Ambulance UK, 26*(6), 324–327.
Burke, M. (2013). Managing work-related stress in the district nursing workplace. *British Journal of Community Nursing, 18*(11), 535–538.
Caroline, N. L. (2007). *Emergency care in the streets*. Burlington: Jones & Bartlett Learning.
Clarke, V., Braun, V., & Hayfield, N. (2015). Thematic analysis. In J. Smith (Ed.), *Qualitative psychology: A practical guide to research methods* (pp. 222–248). Thousand Oaks: Sage.
Clohessy, S., & Ehlers, A. (1999). PTSD symptoms, response to intrusive memories and coping in ambulance service workers. *British Journal of Clinical Psychology, 38*(3), 251–265.
Clompus, S., & Albarran, J. (2016). Exploring the nature of resilience in paramedic practice: A psycho-social study. *International Emergency Nursing, 28*, 1–7.
Cohen, R. L. (2010a). When it pays to be friendly: Employment relationships and emotional labour in hairstyling. *The Sociological Review, 58*(2), 197–218.
Cohen, R. L. (2010b). Rethinking 'mobile work': Boundaries of space, time and social relation in the working lives of mobile hairstylists. *Work, Employment and Society, 24*(1), 65–84.
Davies, C. A. (2008). *Reflexive ethnography: A guide to researching selves and others*. Milton Park: Routledge.
Davis, F. (1959). The cabdriver and his fare: Facets of a fleeting relationship. *American Journal of Sociology, 65*(2), 158–165.
Emerson, R. M., Fretz, R. I., & Shaw, L. L. (2011). *Writing ethnographic fieldnotes*. Chicago: University of Chicago Press.
Eost, C., & Flyte, M. G. (1998). An investigation into the use of the car as a mobile office. *Applied Ergonomics, 29*(5), 383–388.
Filstad, C. (2010). Learning to be a competent paramedic: Emotional management in emotional work. *International Journal of Work Organisation and Emotion, 3*(4), 368–383.
Friese, S. (2019). *Qualitative data analysis with ATLAS*. Thousand Oaks: Sage.
Gabriel, Y. (2000). *Storytelling in organizations: Facts, fictions, and fantasies: Facts, fictions, and fantasies*. Oxford: Oxford University Press.
Goffman, E. (1980). *Behavior in public places: Notes on the social organization of gatherings*. Santa Barbara: Greenwood Press.
Goffman, E. (1990). *The presentation of self in everyday life*. London: Penguin.
Halpern, J., Gurevich, M., Schwartz, B., & Brazeau, P. (2009). What makes an incident critical for ambulance workers? Emotional outcomes and implications for intervention. *Work & Stress, 23*(2), 173–189.
Hammersley, M. (2007). Ethnography. In G. Ritzer (Ed.), *The Blackwell encyclopedia of sociology*. Oxford: Blackwell.
Haraway, D. (1994). A manifesto for cyborgs: Science, technology, and social feminism in the 1980s. In *The Norton anthology of literary criticism*.
Headland, T. N., Pike, K. L., & Harris, M. E. (1990). *Emics and etics: The insider/outsider debate*. Thousand Oaks: Sage.
Herbert, S. (2000). For ethnography. *Progress in Human Geography, 24*(4), 550–568.
Hochschild, A. R. (2012). *The managed heart: Commercialization of human feeling*. Berkeley: University of California Press.
Hollowell, P. G. (1998). *The lorry driver*. New York: Taylor & Francis.
Ikeya, N. (2003). Practical management of mobility: The case of the emergency medical system. *Environment and Planning A, 35*(9), 1547–1564.
Jeffery, R. (1979). Normal rubbish: Deviant patients in casualty departments. *Sociology of Health & Illness, 1*(1), 90–107.
Jennings, K. (2017). Emotional labour in paramedic practice: Student awareness of professional demands. *Journal of Paramedic Practice, 9*(7), 288–294.
Kelly, M. P., & May, D. (1982). Good and bad patients: A review of the literature and a theoretical critique. *Journal of Advanced Nursing, 7*(2), 147–156.

King, E. C., Holliday, P. J., & Andrews, G. J. (2018a). Care challenges in the bathroom: The views of professional care providers working in clients' homes. *Journal of Applied Gerontology, 37*(4), 493–515.

King, N., Horrocks, C., & Brooks, J. (2018b). *Interviews in qualitative research*. Thousand Oaks: Sage.

Lapum, J., Fredericks, S., Beanlands, H., McCay, E., Schwind, J., & Romaniuk, D. (2012a). A cyborg ontology in health care: Traversing into the liminal space between technology and person-centred practice. *Nursing Philosophy, 13*(4), 276–288.

Lapum, J., Ruttonsha, P., Church, K., Yau, T., & David, A. M. (2012b). Employing the arts in research as an analytical tool and dissemination method interpreting experience through the aesthetic. *Qualitative Inquiry, 18*(1), 100–115.

Larsen, J., Axhausen, K. W., & Urry, J. (2006). Geographies of social networks: Meetings, travel and communications. *Mobilities, 1*(2), 261–283.

Laurier, E. (2002a). Notes on dividing the attention of a car driver. *Team Ethno Online, 1*, 1.

Laurier, E. (2002b). The region as a socio-technical accomplishment of mobile workers. In B. Brown & N. Green (Eds.), *Wireless world* (pp. 46–61). New York: Springer.

Laurier, E. (2004). Doing office work on the motorway. *Theory, Culture & Society, 21*(4–5), 261–277.

Maben, J., Adams, M., Peccei, R., Murrells, T., & Robert, G. (2012). 'Poppets and parcels': The links between staff experience of work and acutely ill older peoples' experience of hospital care. *International Journal of Older People Nursing, 7*(2), 83–94.

Madden, R. (2017). *Being ethnographic: A guide to the theory and practice of ethnography*. Thousand Oaks: Sage.

Malone, R. (2003). Distal nursing. *Social Science & Medicine, 56*, 2317–2326.

Marra, M., & Holmes, J. (2002). *Laughing on the inside: Humour and internal politics in the workplace* [Language in the Workplace], Occasional Papers 4. School of Linguistics and Applied Language Studies, University of Wellington.

Mayer, J. D. (1979). Seattle's paramedic program: Geographical distribution, response times, and mortality. *Social Science & Medicine. Part D: Medical Geography, 13*(1), 45–51.

McDonald, S. (2005). Studying actions in context: A qualitative shadowing method for organizational research. *Qualitative Research, 5*(4), 455–473.

McMeekin, P., Gray, J., Ford, G. A., Duckett, J., & Price, C. I. (2014). A comparison of actual versus predicted emergency ambulance journey times using generic Geographic Information System software. *Emergency Medicine Journal, 31*(9), 758–762.

Menzies, I. E. P. (1970). *A functioning of social systems as a defense against anxiety*. London: The Tavistock Institute of Human Relations.

Minney, H., & Bradley, P. (2005). *Taking healthcare to the patient transforming NHS ambulance services*. London: Department of Health.

Mitmansgruber, H., Beck, T. N., & Schüßler, G. (2008). "Mindful helpers": Experiential avoidance, meta-emotions, and emotion regulation in paramedics. *Journal of Research in Personality, 42*(5), 1358–1363.

Mmadi, M. M. (2013). *Mobile workplace: Work conditions and family life of taxi drivers*. Pretoria: University of Pretoria.

Neyland, D. (2007). *Organizational ethnography*. Thousand Oaks: Sage.

Normark, D., & Esbjornsson, M. (2005). *The mobile workplace: Collaboration in a vast setting* (*The interaction society: Practice, theories and supportive technologies*) (pp. 251–270). Hershey: IGI Global.

Ong, M. E., Ng, F. S., Overton, J., Yap, S., Andresen, D., Yong, D. K., et al. (2009). Geographic-time distribution of ambulance calls in Singapore: Utility of geographic information system in ambulance deployment (CARE 3). *Annals Academy of Medicine Singapore, 38*(3), 184.

Orr, J. E. (2016). *Talking about machines: An ethnography of a modern job*. Ithica: Cornell University Press.

Palmer, C. E. (1983). "Trauma junkies" and street work: Occupational behavior of paramedics and emergency medical technicians. *Urban Life, 12*(2), 162–183.
Papper, S. (1970). The undesirable patient. *Journal of Chronic Diseases, 22*(12), 777–779.
Parr, H. (1998a). The politics of methodology in 'post-medical geography': Mental health research and the interview. *Health & Place, 4*(4), 341–353.
Parr, H. (1998b). Mental health, ethnography and the body. *Area, 30*(1), 28–37.
Peleg, K., & Pliskin, J. S. (2004). A geographic information system simulation model of EMS: Reducing ambulance response time. *The American Journal of Emergency Medicine, 22*(3), 164–170.
Pink, S. (2009). *Principles for sensory ethnography: Perception, place, knowing, memory and imagination*. Thousand Oaks: Sage.
Pink, S. (2010). The future of sensory anthropology/the anthropology of the senses. *Social Anthropology, 18*(3), 331–333.
Pink, S. (2015). *Doing sensory ethnography*. Thousand Oaks: Sage.
Psathas, G., & Henslin, J. M. (1967). Dispatched orders and the cab driver: A study of locating activities. *Social Problems, 14*(4), 424–443.
Regehr, C., Goldberg, C., & Hughes, J. (2002). Exposure to human tragedy, empathy, and trauma in ambulance paramedics. *American Journal of Orthopsychiatry, 72*(4), 505–513.
Rowland, E. J. (2014). *Emotional geographies of care work in the NHS*. Unpublished doctoral thesis. London: King's College.
Scott, T. (2007). Sudden traumatic death: Caring for the bereaved. *Trauma, 9*(2), 103–109.
Scott, T. (2013). Sudden death liminality: Dehumanisation and disengagement. *International Emergency Nursing, 21*(1), 10–16.
Steen, E., Næss, A. C., & Steen, P. A. (1997). Paramedics organizational culture and their care for relatives of cardiac arrest victims. *Resuscitation, 34*(1), 57–63.
Stockwell, F., & Stockwell, F. (1972). *The unpopular patient*. London: Royal College of Nursing.
Thompson, J. (1991). Stress in ambulance workers. *Disaster Management, 3*, 193–197.
Tracy, S. J., Myers, K., & Scott, C. (2006). Cracking jokes and crafting selves: Sensemaking and identity management among human service workers. *Communication Monographs, 73*(3), 283–308.
Ueno, N., & Kawatoko, Y. (2003). Technologies making space visible. *Environment and Planning A, 35*(9), 1529–1545.
Urry, J. (2003). Social networks, travel and talk. *The British Journal of Sociology, 54*(2), 155–175.
Uyeno, D. H., & Seeberg, C. (1984). A practical methodology for ambulance location. *Simulation, 43*(2), 79–87.
Van Maanen, J. (2011). *Tales of the field: On writing ethnography*. Chicago: University of Chicago Press.
Verrips, J., & Meyer, B. (2001). Kwaku's car. The struggles and stories of a Ghanian long-distance taxi-driver. In *Car cultures*. Oxford: Berg Publishing.
Watson, D. L. B., Sanoff, R., Mackintosh, J. E., Saver, J. L., Ford, G. A., Price, C., et al. (2012). Evidence from the scene: Paramedic perspectives on involvement in out-of-hospital research. *Annals of Emergency Medicine, 60*(5), 641–650.
Wiberg, M. (2005). Anytime, anywhere in the context of mobile work. In *Encyclopedia of information science and technology* (1st ed., pp. 131–134). Hershey: IGI Global.
Williams, A. (2012a). Emotion work in paramedic practice: The implications for nurse educators. *Nurse Education Today, 32*(4), 368–372.
Williams, A. (2012b). Keeping a "stiff upper lip" in paramedic practice: Coping with emotion work. *Nurse Education Today, 4*(5), 298–298.
Williams, A. (2013a). Hochschild (2003)—The managed heart: The recognition of emotional labour in public service work. *Nurse Education Today, 1*(33), 5–7.
Williams, A. (2013b). The strategies used to deal with emotion work in student paramedic practice. *Nurse Education in Practice, 13*(3), 207–212.

Williams, A. (2013c). A study of emotion work in student paramedic practice. *Nurse Education Today, 33*(5), 512–517.

Williams, P. M., & Shavlik, G. (1979). Geographic patterns and demographic correlates of paramedic runs in San Bernardino, 1977. *Social Science & Medicine. Part D: Medical Geography, 13*(4), 273–279.

Chapter 5
Case Study III: Safe, Ethical Professionals? Trust and the Representation of Nurses, Work and Places in the Context of Neglectful and Dangerous Practice

Elizabeth Peter

Abstract This chapter is an empirical case study based on Elizabeth Peter's original empirical work that showcases contemporary geographies related to professional practice. It presents an analysis of published policy, and institutional, legal and media statements relating to a case where a nurse was found to be an extreme danger to health and life in long-term care. The idea of the chapter is to think about how trust in ethical nursing work, and the places of nursing work, is represented. In particular, attention is paid to the involvement of different scales in work, its regulation and its representation (international, national, city and setting), and how each is used/implicated.

This chapter, the final case study of three, showcases and continues a tradition of researching and articulating 'ethical geographies' in nursing practice; how place, ethics and practice are closely bound and interdependent (Chinn 1997; Liaschenko 1994, 1997, 2001; Liaschenko and Peter 2004; Peter 2002; Peter and Liaschenko 2004), this is itself an empirical development of longstanding ethical/moral human geography (Barnett 2011, 2014; Popke 2006, 2007, 2009; Proctor 1998, 1999; Smith 1997; Valentine 2005). Specifically, we will focus on how trust and trustworthiness are created and enacted in and beyond the institutional setting of the nursing home by focusing on the case of Elizabeth Wettlaufer. Wettlaufer is a Canadian nurse who was found guilty of eight counts of first degree murder, four counts of attempted murder and two counts of aggravated assault by overdosing her patients with insulin from June 2007 to August 2016. Her victims were highly vulnerable people, generally elderly, who were living in long-term care with the exception of one who was receiving nursing care services in her home (CNO 2017). In doing so, we are *not* representing the case of Wettlaufer as an extreme example of unethical nursing behavior because her actions are so unusual that they represent the psychopathology of a healthcare serial killer better than supplying an example to analyze the ethics of a specific nurse or the profession of nursing. When providing reasons

G. J. Andrews et al., *Place and Professional Practice*, Global Perspectives on Health Geography, https://doi.org/10.1007/978-3-030-64179-5_5

for revoking her registration, the College of Nurses of Ontario (2017) stated, "This matter is shameful and unprecedented. It is the most egregious example of abuse and disgraceful conduct that this Panel has ever had to consider" (CNO 2017, p. 6).

We are using this case, because it reveals the trust/mistrust accorded to nurses in general and the places of nursing work. Through the use of victim impact statements written in response to Wettlaufer's crime, we analyze how different scales of nursing work, from the micro to the regulatory and health systems levels, are implicated in how trust is developed and represented. We retain a normative dimension in our analysis to demonstrate the continual growth in interest in how geography relates to ethics and how places can influence ethical and unethical work. In particular, 'relational geographies', as suggested in Chap. 2, acknowledge how ideas, places, things and people are complexly related and dependent on others ideas, places, things and peoples across multiple scales—the co-constitution of phenomenon (Jones 2009; Malpas 2012; Murdoch 2005; Yeung 2005).

Thinking Trust

Trust, trustworthiness and distrust have been the focus of many disciplines, including moral philosophy, bioethics, sociology, political science and medicine (Withers 2018). Trust has also garnered widespread attention in nursing, including two literature reviews on the topic (Dinc and Gastmans 2012, 2013). In geography, however, there has been less attention (e.g. Murphy 2006; Hermes and Poulsen 2013). Withers (2018) has recently stated that "trust should concern human geographers more than it has" (p. 490), arguing that geography like other disciplines in which attention to trust is commonplace is also concerned with justice, equity and well-being. As such, interpersonal trust and trustworthiness have both moral and epistemological dimensions (Withers 2018), because to rely on another requires that we have knowledge of their abilities and can rely on them to have a good will to not hurt us (Baier 1986). A geographical perspective is especially useful, because trust is not only an interpersonal phenomenon; it is also operates within and across social and geographical boundaries (Withers 2018). In institutional settings, it is worth examining how trust and trustworthiness are created and enacted within the placed nature of these settings, and it is also worth reflecting on how trust is inscribed in them, moving across both space and time (Withers 2018).

Trust, understood normatively, has been explored by a number of moral theorists, including feminists. Because feminist theorists view the self as relational, it is not surprising that they have taken great interest in the importance of trust (Baier 1986; Peter and Morgan 2001), examining not only trust issues concerning women but all people, with special attention given to the significance of factors such as gender, class, race and ability. Drawing on the work of moral philosopher Annette Baier (1986), trust can be defined as a reliance on the good will and competence of a person and/or network of persons, such as an institution or group of persons, who has/have been given the responsibility to care for things that another or others value(s). The importance of viewing trust relationships as a network helps to politically and geographically situate relationships of trust because they can be seen as existing on a number of scales. Baier (1986) refers to the "network of trust" (p. 258), "climates of trust" (p. 258) and the "network of relationships" (p. 259). Baier (1994) also refers to "an elaborate chain of trust relationships" (p. 137) and describes how "trust comes in webs" (p. 149). These notions are important, because they acknowledge the complexity of trust relationships and its tendency to ebb and flow.

In terms of a normative theory of trust, a conception of a network is consistent with feminists' tendency to blur the boundaries between the political and the ethical that helps to address complex issues of moral concern. Optimum trusting relationships may not be possible if persons are constrained by forces outside of their immediate relationships such as policies, organizations and social structures. Making the network visible also makes visible the social and geographical boundaries and the dynamic nature of trust. For example, Lawson (2007) has argued that feminist care ethics helps us to be attentive to, not only our local concerns, but also to those that are distant from us given that we are all deeply connected socially by circuits of power and privilege.

The consonance of feminist ethics and recent work on ethics in geography makes it possible to bring the strengths of both together to the examination of trust. For example, Barnett (2014) suggests that normativity, understood geographically, be located directly in the world as an ordinary part of the continual flow of life by focusing on social practices as a means to contextualize individual action. Normativity is implicit in these practices in "which reasons and accounts might be demanded, regrets and disappointments expressed, responsibility taken or disavowed" (Barnett 2014, p. 153). This sense of normativity has similarities with social scientifically oriented accounts of morality as a feature of everyday life (Barnett 2014). In a similar fashion, Walker (1998), a feminist ethicist, describes morality as "a socially embodied medium of mutual understandings and negotiation between people over their responsibility for things open to human care and response" (p. 9). Morality, she states, is located in "practices of responsibility that implement commonly shared understandings about who gets to do what to whom and who is supposed to do what for whom" (Walker 1998, p. 16). Moreover, Walker (1998, 2003), like other feminists, supports the notion that empirical approaches, in the broadest sense, can be used to recognize our moral understandings because they are ultimately created interpersonally. Thus, normativity, in these perspectives from both geography and feminist ethics, is created and embedded in social interactions in which people are called to account for their actions in the 'real' world contexts and places of everyday life.

Feminist ethics also shares many features with relational geography, especially their shared ontological assumptions regarding personhood. Feminist ethicists have critiqued the notion of the purely rational and autonomous person and have instead advocated for a perspective that views persons as unique, gendered, racialized and unequal in power, both rational and emotional, and deeply connected to others on many levels including immediate and distant others (Liaschenko and Peter 2006). Relational geography also recognizes that people exist and are co-created within both immediate and distant networks of people, but adds spatial and material elements that recognize that people's experiences of place have deep affective dimensions and are influenced by other people and objects (Skinner et al. 2015; Andrews et al. 2013). Moreover, spaces and places only come to be through their connections and relationships to other space and places, making them open-ended and networked (Jones 2009; Skinner et al. 2015) and essentially relational in nature.

Methods

This work represents, in general terms, an extreme case study. Case study, as a methodological approach, tends to be used to study a distinct phenomenon in depth, particularly when context plays a central role (Yin 2009). Studying cases has been used extensively in the education of healthcare professionals as a teaching strategy (Stake 2000), including bioethics, and have more recently been used more in empirical research as well. Extreme cases, as opposed to representative cases, such as the one presented in this chapter, have the potential to reveal more knowledge, because they can uncover the deeper roots of a particular problem (Flyvberg 2006). While postpositivism has been the dominant epistemological stance underlying case study methodologies (Yin 2009), critical perspectives, such as poststructuralism, have also been used successfully (Mohammed et al. 2015). In our work, we too have adopted a critical stance in this informal case study in order to capture the elements of power and the particulars of place in our analysis.

The Charges Against Wettlaufer

To begin our analysis, we provide an overview of the events and circumstances surrounding the charges against Wettlaufer. The Superior Court of Justice Agreed Statement of Facts on Guilty Plea of June 1, 2017 details the relevant legal information surrounding the charges against Wettlaufer. Below is a brief summary to give additional context to our case example.

In June 2007, Wettlaufer was hired as registered nurse at Caressant Care Nursing Home in Woodstock, Ontario, Canada. She had access to prescription drugs, including insulin, which was not strictly accounted for or secured at the facility. She often worked nights with little supervision. Her first two victims, Clotilde Adriano and Albina DeMedeiros, were both diabetics whom Wettlaufer gave additional insulin injections to, because she was angry and thought that God wanted them back. They did not die, because they were successfully treated by other nursing staff when they were found to be experiencing low blood sugar. In a similar fashion, she injected nine other residents using excessive doses of insulin that were not prescribed. Seven of these residents died (James Silcox, Maurice Grant, Helen Matheson, Gladys Millard, Mary Zurawinski, Helen Young and Maureen Pickering) and two survived

(Wayne Hedges and Michael Priddle). Later in 2014, Wettlaufer became employed at Meadow Park Nursing Home in London, Ontario where she murdered Arpad Horvath; and subsequently, in 2015 at Telfer Place Long Term Care Facility in Paris, Ontario, she attempted to murder Sandra Towler, again using insulin. Finally, while employed by Saint Elizabeth Health Care in 2016, she attempted to murder Beverly Bertram while providing care to her in Bertram's home. In all situations, she knew what she was doing had no medical purpose, would likely result in death and was being performed without consent for her own gratification alone (Gillese 2019).

In September 2016, Wettlaufer sought help from the Centre for Addiction and Mental Health (CAMH) in Toronto because she was concerned that she could hurt herself or others. It is here, while hospitalized, where she confessed on repeated occasions to her actions, and the police and the College of Nurses of Ontario were contacted by staff. Through the police investigation it was learned that Wettlaufer had confessed her crimes to numerous people previously, but the confessions went unreported, because people did not believe her. For example, between 2009 and 2016, she confessed to her pastor and his wife, a student nurse's aide whom she befriended, a former Narcotics Anonymous Sponsor, a former boyfriend and a former roommate. On June 26, 2017, she was sentenced to life in prison without eligibility for parole for 25 years; and on July 25, 2017, her certificate of registration was revoked by the College of Nurses of Ontario.

Prior to her sentencing, victims of her crimes, including the friends and family of those who died, were given the opportunity to provide victim impact statements (The Canadian Press 2017). It is through these, along with published literature, that we examine how trust, distrust and trustworthiness are produced in the place of the nursing home. Nursing staff at all levels tend to be the focal actors in this regard through their interpersonal work with residents—but also the larger social and political forces which exist beyond the boundaries of the institution, that shape this work must be understood.

Pre-existing Trust and Distrust in Nursing Homes

Several of the victims spoke to the reluctance that they had when making a choice to place their loved one in a nursing home even prior to the Wettlaufer incidents. For example, Donald and Danna Tuck said: "We felt like we were passing her (Maureen Pickering) a sentence of captivity by finding a care home to provide good care"; and Fiona Jane Willis, the daughter of Sandra Towler, said "No one wants to go into a nursing home—no one wants to place a loved one in a nursing home. You hear terrible things about them." A profound societal distrust of nursing homes often exists in Western society, which is likely a reflection of the belief that institutions are dehumanizing and immoral. While homes have been idealized as loving and caring, institutions have been viewed, perhaps sometimes unjustly, as inhumane, rigid, impersonal, abnormal and neglectful (Gleeson and Kearns 2001) and institutional caregiving has been described as routinized and task-based (Banerjee et al. 2012) with staff who are overworked and unable to respond the residents (Holstein et al. 2011). On the one hand, they strive to be homelike as places in which dignity, autonomy and well-being can be maintained (Falk et al. 2013; Holstein et al. 2011). Yet on the other hand, they are institutions that must deliver care to many people often in a limited amount of space and with limited resources. Nursing homes are particularly vulnerable to regimentation, because while they strive to embody the unique qualities of the home, a private place, they also are institutions, public places, representing an assemblage in need of ethical scrutiny (Andrews 2018; Peace and Holland 2001). At times this institutional ethos is a contributing factor to elder abuse, although the prevalence of this abuse in nursing homes and other long-term care facilities is unknown given the lack of reliable evidence available (Pillemer et al. 2016).

The captivity that Donald and Danna Tuck described also suggest that there is an association of the place of the nursing home with a lack of freedom, autonomy and choice. Without a sense of autonomy as emplaced, we cannot adequately recognize how different settings, such as nursing homes, situate autonomy (Andrews and Peter 2006). Because of the efficiencies of institutions, and time and resource restrictions, activities of daily living, such as meals, medication, sleeping, bathing, outings and activities, tend to occur on a schedule that individual residents did not create, significantly constraining autonomy (Banerjee et al. 2012; Mondaca et al. 2018; Palviainen et al. 2003; Powers 2001). It is easy to understand how these types of restrictions to autonomy breed distrust and fear with a sense of the nursing home being more like a prison than a home.

To create a truly homelike environment in which residents can freely exercise their autonomy, it is important that residents have the opportunity to make a world for themselves—a world in which wherever possible they can control decision-making not only over life-sustaining treatments but also over the decisions of everyday life (Holstein et al. 2011). Ideally, the notion of 'placing' someone in a nursing home must be substituted with thinking about them 'living' in the home such that they are making choices about their environment (Reed et al. 2003) and can experience an attachment to and sense of place that is essential for well-being (Wiles et al. 2009; Cheng et al. 2011). This type of autonomy can be best cultivated through care that provides the resources, skill and flexibility of caregivers to focus on individual requests and needs so that they can meet their responsibilities for care. Many workers in nursing homes, however, are faced with the problems of high workloads, poor training, excessive paperwork and a lack of a voice in care (Holstein et al. 2011; Banerjee et al. 2012) that have their roots in policy, regulation and financing that influence the day-to-day activities of the nursing home. While the moral responsibility of nursing staff to provide care that is humanizing and fosters autonomy is in part an interpersonal matter, it can only be realized by the necessary structures of care work that are located across social and geographical boundaries. Unfortunately, even without the actions of a nurse like Wettlaufer, trust in nursing homes and their staff has been weakened by these multiple and interacting social and political forces, even prior to direct experience. These forces only heighten the vulnerability of those who must entrust their well-being, or the well-being of their loved ones, when they are in need of care.

The Betrayal of Trust

The effect of Wettlaufer's actions, as revealed by the victim impact statements, illustrates the additive effect of her actions on the perceived trustworthiness of the nursing home. Elizabeth Silcox (2017), speaking of the murder of her grandfather James Silcox, stated: "I am terrified that someday my dad or his siblings or any one of the people I love may need to go into a nursing home one day. It terrifies me that we cannot trust the institution with our loved one's life. Don't know what I am going to do if the time ever comes that I have to put my father in a home or my children have to put me in a home. The thoughts of this were scary before. Now they are almost debilitating terrifying." Similarly, Andrea Irwin (Silcox), the daughter of James Silcox, stated: "Being a senior part-time employee working in Long Term Care (LTC), I am continually reminded of my father and it breaks my heart knowing that his cries of emotional pain were dealt by someone taking his life and not with compassion as all elderly and LTC residents deserve. LTC facilities are now clouded with distrust and fear. Anyone having the difficult decision of placing their loved one or even themselves in the care of a stranger, has to now fear that this may happen to them as well." The terror of James Silcox's family is clearly evident, displaying their deep affective response in relation to the place of the nursing home and informing their judgment of its lack of trustworthiness (Andrews et al. 2013) Their consequent distrust is understandable given the betrayal of their past trust in the nursing home and their recognition of the inherent vulnerability of their loved ones or themselves who may be in need of care in the future. As Baier (1994) stated, "Bad enough betrayals of trust lead not just to loss of a particular entrusted good but to a lasting inability to partake of that sort of trust-dependent good" (p. 146). In other words, it is not just that the Silcox family will no longer trust one specific nurse or one specific nursing home; their inability to trust anything related to nursing homes or healthcare providers may never be restored.

While trust in healthcare professionals is often idealized and not adequately problematized, it is important that trust, as in the example above, is critically evaluated (Peter and Morgan 2001). Moreover, because trust in relationships is generally invisible until it is violated, examining the breakdown of trust is an indirect way of examining trust. Trust is not inherently good and is not always something we should try to preserve. As Baier (1986) stated, "When the trust relationship itself is corrupt and perpetuates brutality, tyranny or injustice, trusting may be silly self-exposure, and disappointing and betraying trust, including encouraged trust, may be not merely morally permissible but morally praiseworthy" (p. 253). For the victims of Wettlaufer, along with the public, it is not wrong to question the trustworthiness of all dimensions and people associated with nursing home care in order to protect those who are vulnerable. Questioning helps make visible the power differences that can exist among residents, family members and healthcare professionals. These differences may not be unjustified given that frailty and illness can render someone relatively powerless, but they are important to identify because they can sometimes be diminished with advocacy, legal standards and staffing policies (Peter and Morgan 2001).

Self-Trust

Drawing on the work of Baier (1985, 1986), McLeod and Sherwin (2000) and McLeod (2015) have developed the concept of self-trust, arguing that without self-trust people cannot be fully autonomous. Self-trust is needed for people to trust themselves to make good choices that are based on their values, beliefs and goals, and to trust their own judgments so they can act on these decisions. In this way, self-trust is a capacity that is necessary for someone to make autonomous choices that are in their best-interests (McLeod and Sherwin 2000). Self-trust can be undermined by oppression, and also by the experience of betrayal (McLeod 2015) when those betrayed begin to question their ability to make wise judgments.

The experience of the betrayal that erodes self-trust is evident in the statements of the victims. For example, Sharon Young, recalling the death of her aunt Helen Young, stated: "And Beth, you have added insult to injury by recalling in your confession, that I hugged you and thanked you after my Aunt's murder—so not only did I introduce my Aunt to her killer, by deciding to place her at Caressant Care, I also apparently thanked her for her actions. Betrayal doesn't even begin to convey my emotions." Here, Sharon Young is expressing not only her distress of being betrayed, but also alludes to her sense of culpability in her aunt's death, and likely also her distrust of her own judgment through her involvement in placing her aunt at this nursing home.

The manifestation of a loss of self-trust is particularly evident in two of the victim statements. For example, Jon Matheson, the son of Helen Matheson, said: "I placed my Mother in the care of a facility I'd researched never ever considering that she would be the victim of such a despicable act! The question lingers in my head what did I miss or should I have noticed something unusual?" Similarly, the children of Michael Priddle, Jason Priddle and Meredith Maywood stated: "Since news of the abuse our father suffered we have been ruminating on what we missed. When he was agitated was he somehow trying to communicate to us that something was wrong, even though he didn't have the ability to verbalize it. There were many times he appeared anxious and agitated, is this why?" As people trying to meet their moral responsibilities of care to their parents, these unthinkable circumstances have motivated them to take account of their decisions and actions. Their lingering questions represent an agonizing questioning of their own judgment that is the result of the betrayal they have experienced. It is not surprising that people like them experience such injuries to their self-trust because their sense of responsibility and accountability have been brought into question in circumstances they could not reasonably predict or control. It is quite possible that in future they may well have difficulty making decisions in similar circumstances, thereby limiting their autonomy.

Wettlaufer's crimes in terms of both their origins and their impact can only be fully understood through a recognition of the place of the nursing home. While feminist ethics sheds light on the potential impact of betrayal and oppression on the erosion of self-trust, geography can better articulate the complex and emplaced nature of this betrayal that occurred in this case. The multiple layers of actors and places have an intersecting and ever-changing impact not only on the interpersonal nature of trust, but also the intrapersonal nature of trust in the instance of self-trust. Because places possess basic agency (nursing homes provide care, hospitals provide medical treatment and so on), they can influence people, much in the same way that people impact places, whereby 'people make places' and 'places make people' (Andrews 2020, Chap. 3). Recognizing that emotions are central to the embodiment of the experience of place and can shape identity (Skinner et al. 2015; Anderson and Smith 2001) is helpful in understanding how the sense of betrayal of the loved ones of Wettlaufer's victims led to their identities of being capable, caring and responsible adults being damaged and ultimately to the decline of their self-trust.

Multiple Scales of Trust and Distrust in Nursing Homes

Multiple scales of trust and distrust exist in nursing homes beyond those that are interpersonal. Places, understood relationally, are continually produced by many others places, resulting in the care provided being created by structures and ideas that exist on multiple scales (Skinner et al. 2015). The victim impact statements illustrate well how trust and distrust in nursing homes evolved and is multiscalar in nature. Elizabeth Silcox describes the decision to place her grandfather in a nursing home that illustrates these scales well: "Choices made out of love and trust. Trust in a system, trust in a facility, trust in a governing body, trust in an individual, a trust that was ultimately broken in the worst possible way. This murder, this break in trust, this terrible event has turned siblings against each other, broken up our family and caused extreme stress and heart break on all of us... Caused by the cold hearted action of EW (Wettlaufer)." Here Elizabeth Silcox is most likely making reference to the Ontario health care system ("the system"); Caressant Care Nursing Home ("a facility"); the College of Nurses of Ontario, the professional regulator of registered nurses and Wettlaufer ("an individual"). While interpersonal trust and trust in entire organizations are different, they are deeply interrelated (Dinc and Gastmans 2013), revealing the interface between Wettlaufer's actions and intentions and the structures surrounding them, including the profession of nursing, the nursing home and the healthcare system more broadly.

The fragile nature of trust is evident in her words given that this entire network of trust appeared to be broken by the actions of one, presumably malevolent, person. It might be more plausible, however, to recognize that the misdeeds of one person can reveal the checks and balances in a system that potentially could have been able to prevent Wettlaufer from behaving the way she did, such as governing bodies and the nursing home itself. In post-human terms, Wettlaufer's moral agency, and hence her 'choices', may be a reflection that her behaviours could, have been influenced by the place of the nursing home itself in way that she and others could not fully realize or articulate (Andrews 2018). On the one hand, it can be argued that processes in Caressant Care Nursing home should have been able to prevent her actions; but on the other, the structural violence (Banerjee et al. 2012), of the nursing home itself may have played a role in her behavior. In their comparison of Canadian and Scandinavian long-term, residential care, Banerjee et al. (2012) discovered that frontline careworkers in Canada are six times more likely to experience daily physical violence, mainly by residents or their family, than Scandinavian workers. They argued that this violence was frequently normalized "as an inevitable part of eldercare" (Banerjee et al. 2012, p. 390) and attributed this violence to systemic and organizational features influenced by new public management that led to indirect structural violence and poor working conditions. The model of social care in Scandinavia, unlike Canada, is viewed as a State, as opposed to family, responsibility and is embedded in a coordinated national system with strong popular support and more state involvement (Banerjee et al. 2012). It is unclear whether or how this context impacted Wettlaufer, but it is worth consideration. Regardless, however, the devastation to all involved cannot be minimized.

The Nursing Profession

The trust in Wettlaufer prior to her crimes coming to light are not only reflective of her actions and character but also of trust in nurses in general who work across a multitude of settings internationally. Returning to the definition of trust described earlier, this trust can be viewed as being the result of the good will and competence of an entire network of persons. Dianne Crawford, the daughter of James Silcox and a Registered Nurse herself, stated: "I worked diligently as an RN for 40 years to give my patients the best possible care, to enhance their quality of life, to support their families and to be an advocate for their rights….Throughout my career I worked consistently to uphold the dignity and trustworthiness of nursing. Then Elizabeth Wettlaufer destroyed the public trust of the profession." In her statement, it is evident that she expressed her own trustworthiness in two keys ways—by working hard to provide care through not only her knowledge and skill (competence), but also by placing patients' interests first (a good will). In doing so, she was fully aware that she was not only representing herself, but the entire profession.

The 'good will' of nurses, which can be expressed as their honesty and ethical standards, has been rated very highly. Since Gallup (2017) first asked Americans about nurses in 1999, they have been rated as the most ethical and honest profession every year on a list of 21 other professions, with the exception of 2001 when firefighters were placed first on the list after the September 11 terrorist attacks.

Interestingly, in 2017, nursing home operators, who are often nurses, ranked eleventh while members of congress, car salespeople and lobbyists placed 20th to 22nd respectively. It may be that nurses are considered ethical because they have long been influenced by and associated with traditional values and social norms related to gender, specifically discourses surrounding caring (ten Hoeve et al. 2014). This relationship to caring, and perhaps the public's underestimation of nurses' potential power at all levels, may lead the public to believe that nurses are trustworthy because they are kind and harmless. It is through a breach in trust that this power becomes clearly evident. For example, Cheryl Kincaid, the granddaughter Mary Zarawinkski, stated: "This criminal has weakened trust in the professionals that we put our faith in to look after our aging family members. We fear, that had she not confessed, the power of this nurse had and her ability to execute her plans, killing and nursing many vulnerable people without being caught, has weakened our faith in the profession and the senior care facilities." The Public Inquiry into the Safety and Security of Residents in the Long-Term Care Home System (Gillese 2019) that followed the Wettlaufer offences confirms that without Wettlaufer's confession her offences would not have come to light.

Although not evident in the witness impact statements, nurses are not always trusted as skilled and knowledgeable professionals (i.e. their competence is not always trusted or is not fully recognized). ten Hoeve et al. (2014) argue that, despite the significant increase in educational attainment and knowledge production in nursing, the public does not often recognize nurses as highly skilled professionals. In part, this is the result of the long-standing subordination of nursing to medicine (ten Hoeve et al. 2014) that may be the result of medicine's early professionalization that allowed physicians to secure boundaries around domains of knowledge and attain a high level of social and economic rewards in comparison with other groups, such as nursing, that professionalized later (Fox and Reeves 2015). Here again, when people trust, or do not trust, individual nurses like Wettlaufer, they are trusting a multiscalar network of nurses and other health care professionals. They also are not only trusting in the ethical nature of a profession but its presumed knowledge and expertise as well.

Trust and distrust in the international network of nurses also involves trust and distrust in the places where nurses are educated and employed. For example, nurses' push for higher education can be in part viewed as a mechanism to be associated with the status and trust in institutions of higher education, such as universities and colleges. These institutions also tend to foster the intellectual work, as opposed the 'dirty' body work, of nurses (Parker 1997) creating motivation for nurses to move away from the bedside to assume other roles in management and education. Nurses' relationship with hospitals also plays a role in society's trust in them. While historically hospitals were associated domestic functions and the care of the poor and contagious, today they are associated with the prestige of high technology and medicine (Risse 1999). Nursing homes, in comparison, do not hold this kind of prestige and presumably would lower the trust in the people who work in them. They also tend to be staffed with a variety of less educated aids and a smaller number of registered nurses (Armstrong and Braedley 2013).

Regulation of Nursing

Like other healthcare professions, nursing is self-regulated in order to protect the public and to ensure quality patient care. Legislation requires regulatory bodies to establish and monitor practice standards and codes of ethics in order to retain the trust of the public (ICN 2013). Self-regulation is regarded as a privilege that recognizes the knowledge and expertise of a profession and its ability to place the public's interests above its own professional interests (CNO 2018). Regulators normally enact their responsibilities as distant actors whose responsibility is to maintain the trustworthiness of professions and whose impact is the co-creation of professionals, such as nurses. At the time of her crimes, Wettlaufer was registered with the College of Nurses of Ontario, which is the governing body for Registered Nurses in Ontario, Canada that establishes entry to practice requirements, articulates and promotes practice standards, administers a quality assurance program and enforces standards of practice and conduct (CNO 2018). The College works in conjunction with educational institutions, employers and government (CNO 2018), illustrating multiple scales of trust with respect to regulation. In fact, her disciplinary hearing decision states: "It is a privilege for members of this College to be the guardians of the public trust. The heinous actions by this one member have violated the public's confidence in the nursing profession and placed a stain on the many dedicated professionals who care for their clients with knowledge and compassion" (CNO 2017, p. 7).

The role of regulation was evident in the statement of Dianne Crawford, likely relying on her experience as a Registered Nurse, who stated: "I felt anger and disappointment: anger that this sort of thing can happen in Ontario a province where we are proud of our health care system, and disappointment that so many employers refuse to admit to the College of Nurses or the police that a staff member has abused a resident, stolen medication, made serious medication errors. Instead, those same staff members are often quietly discharged, only to get a job at another facility and often continue their illegal, unprofessional actions." From her perspective, the employer, in this case Caressant Care Nursing, had a responsibility to report Wettlaufer to the College as opposed to allowing her to move on to a new employer without the new employer being aware of her history. While Wettlaufer's employment record described 44 instances of medication errors, disciplinary actions and warnings for incompetence, only 10 of these were reported to the College when she was fired in March 2014 from the Caressant Care in March 2014 (Contenta 2018). Without this kind of reporting, the College could not play the part it needed to in this intricate network of trust that was necessary to protect Wettlaufer's patients. As such, normativity, as Barnett (2014) describes it, exists directly in the nursing home with its expected social practices to make reports to regulators. These practices, informed by regulation, contextualize the actions of both administrators in the home and its nurses.

It is important to recognize that the Public Inquiry (Gillese 2019) did not find any individual misconduct and praised the dedication of the majority of those working in long-term care. While individual shortcomings were identified, Wettlaufer's actions were attributed to systematic vulnerabilities that were described as "complex, multifaceted, and polycentric in nature" (Gillese 2019, p. 14), reflecting weaknesses that can be understood and addressed by recognizing that individual actions are embedded in larger networks that exist on multiple scales.

The Health Care System

At the broadest and most geographically dispersed level, health care systems with their large array of hospitals, community health centers, private homes, services, regulators, legislation, workers, regulators, organizations and so forth are in constant flux, influencing nursing homes and our trust/distrust in them. It is not surprising that those who live in nursing homes in Canada are mainly seniors. In 2011, 7.1% of seniors lived in some form of a collective dwelling that provided specialized care, including nursing homes, long-term care hospitals and seniors' residences (Statistics Canada 2018). Expectedly, older seniors are more likely to live in one of these specialized facilities within the health care system, with 29.6% of seniors aged 85 and older living in one in Canada in 2011 (Statistics Canada 2018). Workers in these settings tend to be of lesser status and lower paid than those in the acute care sector, with many workers often being female, immigrants or from racialized communities (Armstrong and Braedley 2013). The characteristics of this workforce can be viewed as a reflection of the public's valuing of the acute care sector of the healthcare system above that of chronic and social care sectors. Despite the growing of acuity and complexity of healthcare needs of these residents, the work is often assumed to be relatively unskilled and an extension of women's domestic labour, with the limited power and remuneration of these workers reflected in these assumptions (Armstrong and Braedley 2013). Given the current neoliberal liberal climate, needs not met by paid caregivers, which are many, must be fulfilled by volunteers, family and friends (Armstrong and Braedley 2013).

It is within this climate that regulations, which guide the quality and safety of care provided, take force. While these demand the quality and dignity of care for residents, limited resources necessitate that these regulations do not overly restrict care providers and do not set the quality standard too high in terms of staffing and other amenities. Reference to the Canadian health care system and the place of the nursing home within it was evident in the victim impact statements. For example, Dianne Crawford said: "I know from working in hospitals and nursing homes that it is easy to access both insulin and syringes…I wonder if greater control over those medications could have saved my father's (James Silcox) life—I just don't know. Where do I stand? Do I condemn a system that has worked well in hospitals and nursing homes for years? My trust in a system I believed was basically sound has been shattered." Implicitly, she was questioning whether the regulations that give nurses, like herself, the freedom of access to powerful medications such as insulin should be examined given her father's death. The entire taken-for-granted network of trust becomes suspect given the freedoms within it that normally permit the provision of good care, but in the case of Wettlaufer made it possible for her to murder people without difficulty or detection.

One aspect of this assumed trust in this situation relates to attitudes and beliefs related to Canada's healthcare system. Dianne Crawford's specific reference to pride in Ontario's health care system locates her trust in a very specific way in that she was not trusting health care in general, or everywhere and anywhere. Universal health care in Canada tends to form an important aspect of national identity, particularly in English speaking regions, with many Canadians taking great pride in it (Dufresne et al. 2014) Access to healthcare also is imbued with normative dimensions with healthcare services being viewed by most as a human right, as opposed to a commodity, even with the growing pressures of neoliberalism.

Conclusions

The statements made by Wettlaufer's victims have revealed how trust is created and destroyed in the setting of the nursing home through multiple scales of nursing and healthcare work. Through the examination of these statements, the insights of both geographers and feminist ethicists were found to be complementary perspectives that can make possible an analysis of the normative and spatial dimensions of practices in nursing homes, particularly the nature of trust and distrust in healthcare practices. This work represents one of the first in nursing to incorporate relational geography as a theoretical lens, serving as a potential example to further develop geographical scholarship in nursing. While the development and application of geographical insights have a long history in nursing scholarship, places have tended to be conceptualized as stable with little attention given to relationships and places that exist outside of them that a relational perspective could remedy. The use of relational geography in nursing, therefore, could provide the necessary turn in nursing scholarship to refresh not only nursing geography but nursing ethics as well. Empirical applications are potentially numerous. At the time of writing the COVID-19 pandemic, and the crisis it has meant for nursing homes and their residents, is a situation which is crying out for such a relational approach. Other situations will surely arise in future.

References

Anderson, K., & Smith, S. J. (2001). Emotional geographies. *Transactions of the Institute of British Geographers, 26*(1), 7–10.

Andrews, G. J. (2018). *Non-representational theory & health: The health in life in space-time revealing*. Milton Park: Routledge.

Andrews, G. J. (2020). Health geographies III: More-than-representational pushes and expressions. *Progress in Human Geography, 44*(5), 991–1003.

Andrews, G. J., Evans, J., & McAlister, S. (2013). 'Creating the right therapy vibe': Relational performances in holistic medicine. *Social Science & Medicine, 83*, 99–109.

Andrews, G. J., & Peter, E. (2006). Moral geographies of restraint in nursing homes. *Worldviews on Evidence-Based Nursing, 3*(1), 2–7.

Armstrong, P., & Braedley, S. (2013). Introduction. In P. Armstrong & S. Braedley (Eds.), *Troubling care: Critical perspectives on research and practices* (pp. 9–16). Toronto: Canadian Scholars' Press.

Baier, A. (1985). Postures of the mind: Essays on mind and morals. U of Minnesota Press.

Baier, A. (1986). Trust and antitrust. *Ethics, 96*, 231–260.

Baier, A. (1994). *Moral prejudices: Essay on ethics*. Cambridge: Harvard University Press.

Banerjee, A., Daly, T., Armstrong, P., Szebehely, M., & Armstrong, H. (2012). Structural violence in long-term, residential care for older people: Comparing Canada and Scandinavia. *Social Science & Medicine, 74*, 390–398.

Barnett, C. (2011). Geography and ethics: Justice unbound. *Progress in Human Geography, 35*(2), 246–255.

Barnett, C. (2014). Geography and ethics III: From moral geographies to geographies of worth. *Progress in Human Geography, 38*(1), 151–160.

Cheng, Y., Rosenberg, M. W., Wang, W., Yang, L., & Li, H. (2011). Aging, health and place in residential care facilities in Beijing, China. *Social Science & Medicine, 72*, 365–372.

Chinn, P. L. (1997). Response to "Ethics and the geography of the nurse-patient relationship: Spatial vulnerable and gendered space". *Research and Theory for Nursing Practice, 11*(1), 61.

College of Nurses of Ontario (CNO). (2017, July 25). Discipline Committee of the College of Nurses of Ontario: CNO vs. Elizabeth Tracy Mae Wettlaufer Registration # 9581737. Heard.

College of Nurses of Ontario (CNO). (2018). About the College of Nurses of Ontario. Retrieved from http://www.cno.org/en/what-is-cno/

Contenta, S. (2018). Much of Elizabeth Wettlaufer's stained record not reported to College of Nurses. Retrieved from https://www.thestar.com/news/canada/2018/06/11/nurse-who-killed-8-patients-faced-discipline-at-care-home-inquiry-hears.html

Dinc, L., & Gastmans, C. (2012). Trust and trustworthiness in nursing: An argument-based literature review. *Nursing Inquiry, 19*, 223–237.

Dinc, L., & Gastmans, C. (2013). Trust in nurse-patient relationships: A literature review. *Nursing Ethics, 20*(5), 501–516.

Dufresne, Y., Jeram, S., & Pelletier, A. (2014). The true north strong and free healthcare? Nationalism and attitudes towards private healthcare options in Canada. *Canadian Journal of Political Science, 47*(3), 569–595.

Falk, H., Wijk, H., Persson, L., & Falk, K. (2013). A sense of home in residential care. *Scandinavian Journal of Caring, 27*(4), 999–1009.

Flyvberg, B. (2006). Five misunderstandings about case-study research. *Qualitative Inquiry, 12*(2), 219–245.

Fox, A., & Reeves, S. (2015). Interprofessional collaborative patient-centred care: A critical exploration of two related discourses. *Journal of Interprofessional Care, 29*(2), 113–118.

Gallup (2017) Poll on nursing. Gallup

Gillese, E. E. (2019). Public inquiry into the safety and security of residents in the long-term care homes system. Toronto, ON: Commission d'enquête publique sur les foyers de soins de longue durée.

Gleeson, B., & Kearns, R. (2001). Remoralising landscapes of care. *Environment and planning D: Society and Space, 19*(1), 61–80.

Hermes, K., & Poulsen, M. (2013). The intraurban geography of generalised trust in Sydney. *Environment and Planning A, 45*(2), 276–294.

Holstein, M. B., Parks, J. A., & Waymack, M. H. (2011). *Ethics, aging, and society. The critical turn*. New York: Springer.

International Council of Nurses. (2013). Position statement: Nursing regulation. Retrieved from https://www.icn.ch/sites/default/files/inline-files/B04_Nsg_Regulation.pdf

Jones, M. (2009). Phase space: Geography, relational thinking, and beyond. *Progress in Human Geography, 33*(4), 487–506.

Lawson, V. (2007). Geographies of care and responsibility. *Annals of the Association of American Geographers, 97*(1), 1–11.

Liaschenko, J. (1994). The moral geography of home care. *Advances in Nursing Science, 17*, 16–25.

Liaschenko, J. (1997). Ethics and the geography of the nurse–patient relationship: Spatial vulnerabilities and gendered space. *Scholarly Inquiry for Nursing Practice, 11*, 45–59.

Liaschenko, J. (2001). Nursing work, housekeeping issues, and the moral geography of home care. In D. N. Weisstub, D. C. Thomasma, S. Gauthier, & G. F. Tomossy (Eds.), *Aging: Caring for our elders* (pp. 123–137). New York: Springer.

Liaschenko, J., & Peter, E. (2004). Nursing ethics and conceptualizations of nursing: Profession, practice and work. *Journal of Advanced Nursing, 46*(5), 488–495.

Liaschenko, J., & Peter, E. (2006). Feminist ethics: A way of doing ethics. In A. Davis, V. Tschudin, & L. de Raeve (Eds.), *Essentials of teaching and learning in nursing ethics: Content and methods* (pp. 181–190). London: Elsevier.

Malpas, J. (2012). Putting space in place: Philosophical topography and relational geography. *Environment and Planning D: Society and Space, 30*(2), 226–242.

McLeod, C. (2015). *Trust* (Stanford encyclopedia of philosophy. Center for the study of language and information). Stanford: Stanford University.

McLeod, C., & Sherwin, S. (2000). Relational autonomy, self-trust and health care for patients who are oppressed. In C. Mackenzie & N. Stoljar (Eds.), *Relational autonomy: Feminist perspectives on autonomy, agency and the social self* (pp. 259–279). Oxford: Oxford University Press.

Mohammed, S., Peter, E., Gastaldo, D., & Howell, D. (2015). Rethinking case study methodology in poststructural research. *Canadian Journal of Nursing Research, 47(1)*, 97–114.

Mondaca, M., Josephsson, S., Katz, A., & Rosenberg, L. (2018). Influencing everyday activities in a nursing home setting: A call for ethical and responsive engagement. *Nursing Inquiry, 25*(2), e12217.

Murdoch, J. (2005). *Post-structuralist geography: A guide to relational space*. Thousand Oaks: Sage.

Murphy, J. T. (2006). Building trust in economic space. *Progress in Human Geography, 30*(4), 427–450.

Palviainen, P., Hietala, M., Routasalo, P., Suominen, T., & Hulpi, M. (2003). Do nurses exercise power in basic care situations? *Nursing Ethics, 10*(3), 269–280.

Parker, J. M. (1997). Searching for the body in nursing. In J. Lawler (Ed.), *The body in nursing: A collection of views*. Melbourne: Churchill.

Peace, S., & Holland, C. (2001). Homely residential care: A contradiction in terms? *Journal of Social Policy, 30*(3), 393–410.

Peter, E. (2002). The history of nursing in the home: Revealing the significance of place in the expression of moral agency. *Nursing Inquiry, 9*(2), 65–72.

Peter, E., & Liaschenko, J. (2004). Paradoxes of proximity: A spatio-temporal analysis of moral distress and moral ambiguity. *Nursing Inquiry, 11*(4), 218–225.

Peter, E., & Morgan, K. (2001). Explorations of a trust approach for nursing ethics. *Nursing Inquiry, 8*(1), 3–10.

Pillemer, K., Burnes, D., Riffin, C., & Lachs, M. S. (2016). Elder abuse: Global situation, risk factors, and prevention strategies. *Gerontologist, 56*(S2), S194–S205.

Popke, J. (2006). Geography and ethics: Everyday mediations through care and consumption. *Progress in Human Geography, 30*(4), 504–512.

Popke, J. (2007). Geography and ethics: Spaces of cosmopolitan responsibility. *Progress in Human Geography, 31*(4), 509–518.

Popke, J. (2009). Geography and ethics: Non-representational encounters, collective responsibility and economic difference. *Progress in Human Geography, 33*(1), 81–90.

Powers, B. A. (2001). Ethnographic analysis of everyday ethics in the care of nursing home residents with dementia: A taxonomy. *Nursing Research, 50*(6), 332–339.

Proctor, J. (1999). *Geography and ethics: Journeys in a moral terrain*. Hove: Psychology Press.

Proctor, J. D. (1998). Ethics in geography: Giving moral form to the geographical imagination. *Area, 30*(1), 8–18.

Reed, J., Cook, G., Sullivan, A., & Burridge, C. (2003). Making a move: Care home residents experiences of relocation. *Ageing and Society, 23*, 225–241.

Risse, G. B. (1999). *Mending bodies, saving souls: The history of hospitals*. Oxford: Oxford University Press.

Skinner, M. W., Cloutier, D., & Andrews, G. J. (2015). Geographies of ageing: Progress and possibilities after two decades of change. *Progress in Human Geography, 39*(6), 776–799.

Smith, D. M. (1997). Geography and ethics: A moral turn? *Progress in Human Geography, 21*(4), 583–590.

Stake, R. (2000). Case studies. In N. Denzin & Y. Lincoln (Eds.), *Handbook of qualitative research* (pp. 435–454). Thousand Oaks: Sage.

Statistics Canada. (2018). Living arrangements of seniors. Retrieved December 17, 2018 from https://www12.statcan.gc.ca/census-recensement/2011/as-sa/98-312-x/98-312-x2011003_4-eng.cfm

Superior Court of Justice (Southwest Region at Woodstock). (2017). Her majesty the Queen and Elizabeth Tracy Mae Wettlaufer agreed statement of facts on Guilty Plea. Retrieved from https://blackburnnews.com/wp-content/uploads/2017/06/Agreed-statement-of-fact-Wettlaufer-FINAL.pdf

ten Hoeve, Y., Jansen, G., & Roodbol, P. (2014). The nursing profession: Public image, self-concept and professional identity. A discussion paper. *Journal of Advanced Nursing, 70*(2), 295–309.

The Canadian Press. (2017). *Wettlaufer victim impact statements*. San Francisco: Scribd. Retrieved from https://www.scribd.com/document/352289376/Wettlaufer-Victim-Impact-Statements.

Valentine, G. (2005). Geography and ethics: Moral geographies? Ethical commitment in research and teaching. *Progress in Human Geography, 29*(4), 483–487.

Walker, M. U. (1998). *Moral understandings: A feminist study in ethics*. New York: Routledge.

Walker, M. U. (2003). Truth and voice in women's rights. Marquette University E-Publications.

Wiles, J. L., Allen, R. E. S., Palmer, A. J., Hayman, K. J., Keeling, S., & Kerse, N. (2009). Older people, their social spaces an article of wellbeing, attachment to place in Zealand Aotearoa New. *Social Science & Medicine, 68*, 664–671.

Withers, C. W. J. (2018). Trust—In geography. *Progress in Human Geography, 42*(4), 489–508.

Yeung, H. W. C. (2005). Rethinking relational economic geography. *Transactions of the Institute of British Geographers, 30*(1), 37–51.

Yin, R. (2009). *Case study research: Design and methods* (4th ed.). Thousand Oaks: Sage.

Part III
Visions

Chapter 6
Towards a Research Agenda That Progresses Key Debates: Example I—Unpacking More-Than-Human Assemblages of Person-Centred Care

Gavin J. Andrews, Emma Rowland, and Elizabeth Peter

Abstract This chapter explores the potential for a more integrated and central geographical contribution to mainstream professional practice debates. As an example it describes how geography might play a fuller and more embedded role in understandings of 'person-centred care' (PCC). After reviewing traditional research areas on PCC, it describes how PCC might be understood as a 'more-than–human' spatial production. Specifically it describes how a posthumanist theoretical orientation might recast some humanistic concepts central in PCC (identity and meaning, emotions, difference and oppression, agency and communication). Then, using a three-part posthumanist typology, it considers how PCC might; (1) emerge and express within material social assemblages; (2) be enacted and performed affectively by vital bodies and vibrant objects; (3) be conducted in immediate, pre-personal, more-than-representational spacetimes.

Introduction

This chapter showcases an area in which geography might play a fuller and more embedded role, that being understandings of 'person-centred care' (PCC). After reviewing four traditional research areas on PCC, the chapter describes how PCC might be understood as a 'more-than-human' spatial production. It outlines how a posthumanist theoretical orientation might rethink some humanistic concepts important in PCC (identity and meaning, emotions, difference and oppression, agency and communication). Thereafter, using a recently developed, posthumanist typology, it considers how PCC might (1) emerge and express within material social assemblages; (2) be enacted and performed affectively by vital bodies and vibrant objects; and (3) be conducted in immediate, pre-personal, more-than-representational spacetimes. The chapter concludes with some thoughts on how these processes might be unpacked in research.

G. J. Andrews et al., *Place and Professional Practice*, Global Perspectives on Health Geography, https://doi.org/10.1007/978-3-030-64179-5_6

PCC: Development and Key Facets

PCC is a model of care that has emerged rapidly in the past two decades. Quickly becoming synonymous with 'best practice', a rush has occurred amongst policy makers, and academic and health care institutions, to claim PCC; it being deployed across a full range of professional roles and clinical specialisms (Brooker 2003). A good part of the literature on PCC describes its history and key facets. As Mitchell and Agnelli (2015) note, PCC is not a new concept, its principles being traceable back to the work of the humanistic psychologist Carl Rogers in the 1950s and 1960s and his ideas on accounting for personhood in therapy (Rogers 1951, 1959, 1961). PCC's origins can also be traced to nursing with the work of Abdellah et al. (1960) and Levine (1967) who emphasized the individualization of the patient and their care. Levine (1967) warned, however, that the term patient-centred nursing could too easily become a 'wordfact' (p. 53), meaning that a word or term that could too easily become a substitute for a reality. Moreover, PCC's origins can also be loosely traced to certain parts of humanistic nursing models and theories—or humanistic parts of general nursing models and theories—of approximately the same period, such as how the patient is conceived in Peplau's theory of 'interpersonal relationships in care' (Peplau 1952a, b), or how sympathy and empathy is arrived at in Travelbee's 'human-to-human relationship model' (Travelbee 1971). Later in the 1990s PCC was put on the map by the work of Thomas Kitwood specifically on dementia care (Kitwood 1997). Indeed, Kitwood described how a 'malignant social psychology' related to people with dementia is active in society due to their invisibility, this resulting in a range of negative attitudes and behaviours that undermine their personhood and wellbeing. In particular Kitwood described many ways in which the personhood of clients with dementia may be undermined in care settings. He argued that depersonalising tendencies often occur because of a lack of understanding amongst healthcare professionals as to the nature and needs of clients (rather than through malicious intent) and argued how the unique person and their experience should be prioritised, as well as an attentiveness paid to each unique encounter.

In more recent years, development of PCC has emerged partly as a response to criticisms of professional work, and the erosion of public confidence, following high profile cases of mistreatment and dangerous practice; it is part of a move to ensure that safe high-quality health care will be provided in future (Ross et al. 2015). Theoretically more recent years have also involved the articulation of some of the core components/principles of PCC (see Cloninger 2011; Kitson et al. 2013; Sharma et al. 2015). As Brooker (2003) suggests, no one single definition of these—or indeed PCC more generally—has emerged; definitions differing according to the clinical specialities and contexts in which they might be applied. Moreover, whilst some have emphasized clients, others have emphasized values in practice, and others have applied aspects of practice. With a philosophical emphasis, McCormack (2004), for example, takes a philosophical approach to the person and to the nature/attributes of human being. Specifically, he starts by considering the things about humans that are most fundamental, intrinsic and distinguish us from other biological life (including self awareness, second level analytical thinking, placing non-utility worth in others, dignity). From this, his 'four pillars of PCC' that he believes raise the importance of knowing the person/biography, of relationships, and of seeing behind immediate needs are:

1. *Being in relation.* A realization that persons exist in relationships with and relationally to other persons.
2. *Being in the social world.* A realization that persons are social networked beings.
3. *Being in place.* A realization that persons have spatial contexts through which their personhood is articulated.
4. *Being with self.* A realization that being recognized and respected impacts on a person's sense of self and worth.

Otherwise, with an ethical emphasis on values in practice, Brooker (2003) articulates four broadly accepted components of PCC (derived specifically from their emphasis on dementia care):

1. *Fully valuing both clients and caregivers.* A realization that all clients are agents regardless of their mental or physical status, and a realization that stereotyping should be resisted and anti-discriminatory practice enhanced.
2. *Treating all humans as individuals.* A realization that all people are unique and have unique needs. In turn this involving an acknowledgement that individual assessment is critical.
3. *Attempting to look at situations through the eyes of the client.* A priority to know the client's reality and perspective and to cater to their viewpoint.
4. *Working towards a positive social environment within which clients wellbeing is maximised.* A professional therapeutic alliance, including networked clinical relationships and day-to-day interactions (not just specific interventions) and interpersonal skills development.

More practically with an emphasis on applied aspects of practice, Manley et al. (2011), for example, articulate seven priorities of PCC:

1. *Getting to know clients as full persons.* This including their values, beliefs, aspirations, needs and preferences.
2. *Enabling clients to make decisions.* This based on options available to them.
3. *Shared decision making between clients and professionals.* This to maximise choice.
4. *Providing information tailored to each client.* This to assist them in making decisions based on the best evidence available.
5. *Assisting clients in interpret technical information and evidence.* This helping them to understand their options and the consequences of each.
6. *Supporting clients to assert their choices.* This might be direct or involve supporting an appointed advocate to represent these choices.
7. *Conducting ongoing evaluation.* This including collecting feedback to insure that care continues to be appropriate and maximised for each client.

PCC: Current Implementation and Practice Issues

Another area of literature is focused broadly on the current state of the art of PCC, and part of it pays attention to cultures of care and to some of the hurdles in the way of achieving full PCC. Scholars have highlighting diverse issues, including overcoming routinized non-PCC behaviours and resistance to change; sustaining commitment to PCC; ensuring consistency beyond adhoc or occasional person centered people or moments; developing PCC as a specific omnipresent culture; ongoing institutional development of what PCC is; and the need for strong leadership and comprehensive reform at a 'whole system' level (Eaton et al. 2015; McCormack et al. 2011; Carlström and Ekman 2012).

A second part of the literature is concerned with the numerous specificities and practicalities in achieving PCC, including focuses on staff understanding of what PCC is (Ross et al. 2015); dealing with scepticism and stereotypical attitudes in staff (Moore et al. 2017); enabling attitudes from leaders (Moore et al. 2017); ensuring the active role of clients for their own empowerment (Broderick and Coffey 2013; Moore et al. 2017; Sharp et al. 2016); dealing with incomplete records (Broderick and Coffey 2013); translating classroom theory into teams and settings and ensuring learning opportunities (McCarthy 2006; Skaalvik et al. 2010); ensuring ethical practice in PCC such as confidentiality (Munthe et al. 2012); taking all opportunities available for PCC (Clissett et al. 2013); melding PCC with a recovery ethos and goals (Hill et al. 2010); developing interpersonal skills (Doherty and Thompson 2014); resisting clinical information systems that function to impede nurses' awareness of patients' subjectivities (Rankin 2015); and establishing phases and routines in PCC relationships (Ekman et al. 2011). Notably, in certain studies, the practice and work environment are deemed to be particularly important. Sjögren et al. (2015), for example, find a range of factors to be key to PCC, including high levels of staff satisfaction, low role stress, supportive unit culture and higher proportions of staff with higher education qualifications (see also Edvardsson et al. 2008).

A third and final part of this literature is focused on outcomes, such as Finset (2011), who found PCC to have positive effects on patient satisfaction, patient adherence, health care utilization, malpractice litigation and health outcomes. Meanwhile, other studies develop and describe tools that are important to measuring and insuring PCC. Davis et al. (2008), for example, develop a tool to record client experience, and Rokstad et al. (2012) develop a tool for staff self-reporting and rating of PCC.

PCC: Programs, Models and Frameworks

A broad area of literature is concerned with practice development in PCC through particular programs/interventions and models/frameworks. With regard to programs, these are necessarily focused yet they provide valuable data on what is specifically required in particular contexts to achieve PCC. McCance et al. (2013), for example, explored a facilitated program in acute care settings and found that success hinges upon good working relationships and prioritization. McCormack et al. (2010) evaluated residential settings for older people using an index to measure change finding teamwork, workload management and staff interrelationships to be key to success. Kontos et al. (2010) implemented a drama program in dementia care, finding that it conveyed meaning and increased efficiency in care and independence in clients. Ekman et al. (2011) oversaw and intervened in cardiac wards finding that PCC shortened hospital stays.

Randomised control trials (RCTs) have been used at times to evaluate programs. Ballard and Aarsland (2009), for example, used an RCT to evaluate PCC through dementia care mapping in dementia contexts, finding it to result in fewer falls than others forms of care. Brännström and Boman (2014) also used an RCT in heart failure palliative contexts finding PCC to improve quality of life and morbidity. Fors et al. (2015) used an RCT for after care following cardia event. They found that PCC increased self efficacy but found no other clinical improvement. On the whole, however, survey or qualitative research remains the norm for evaluating PCC, largely due to the important experiential and personal elements and the subtleties in practice and outcomes.

With regard to models/frameworks, these have been plentiful and have proved useful in assisting implementation and practice development in PCC. Most notably, McCormack and McCance (2006) have developed a broad framework for PCC in nursing, generally composed of four concepts: prerequisites (attributes of the nurse), the care environment (context in which care is delivered), process (delivery through activities), and outcomes (results through measures). Other important contributions with regard to models have included Britten et al. (2017) on the 'Gothenburg model', Alharbi et al. (2014) on models from the patient's perspective, Parley (2001) on measuring model outcomes, and Santana et al. (2018) on technology, monitoring and reporting.

PCC: Emerging Understandings and Extensions

A final area of literature conveys the latest cutting edge thinking and knowledge on PCC and beyond, some of which can be unconventional or unexpected. O'Rourke et al. (2019), for example, argue that PCC be reconceptualized to focus on interdependence as opposed to independence, emphasizing shared decision-making and mutual responsibility and accountability in the relationship between patient and provider. Along the same lines, Nolan et al. (2004) think beyond PCC in gerontological nursing to 'Relationship Centered Care' (RCC) that, they feel, would deal more with symmetry in relationships and with networks beyond (such as families). Relationships, they argue, make people feel part of things, providing continuity, purpose, achievement and significance. Similarly, Dewar and Nolan (2013) talk about RCC in long term care settings as a way to aid professionals delivering care in practice based on 'appreciative caring conversations'. These enable all parties to obtain forms of 'person and relational knowledge' about 'who people are and what matters to them'.

Kontos (2005) similarly thinks beyond PCC. She argues that dementia care should incorporate the idea that the body is a fundamental source of selfhood that does not necessarily derive its agency from full-cognition. Hence, she advances the idea of 'embodied selfhood'. In terms of implications for practice, Kontos argues that embodied selfhood is something that must be integrated in many caring interactions and in the way clients are regarded. Carers, she notes, need to acknowledge how the self resides in the pre-reflective body, which has a performative coherence. Hence, subtle attentiveness on the part of carers to appearance, cleanliness, etiquette, bodily expressions, gestures and spontaneity are all important. Developing these ideas, Kontos and Naglie (2007) talk about imagination and the body in PCC. Specifically, the importance of finding a 'sympathetic connection' through imagination (placing oneself in the shoes of the client). Thus, imagination bridges the gap between the care provider and care recipient, enabling the former, with their new compassionate disposition, to act with greater compassion.

Posthumanism: An Introduction

Posthumanism has never been a universal, cohesive or totally consistent movement, incorporating a number of theoretical orientations each possessing its own epistemological, methodological and empirical specificity (Castree et al. 2004a, b). These theoretical orientations have included new materialisms (Whatmore 2006), assemblage theory (Anderson and McFarlane 2011), relational thinking (Jones 2009), actor network theory (Murdoch 1998), non-representational theory (Thrift 2008), affect theory (Pile 2010), vitalist philosophy (Philo 2007) and post-phenomenology (Ash and Simpson 2016). Still, despite the differences between these theoretical orientations, there still are general harmonies, commonalities and compatibilities. For a start, as Braidotti (2019a) argues, they have comfortably co-existed in and co-facilitated the emergence of a number of transdisciplinary 'hubs'—creative yet applied empirical convergences of posthumanism. These include biosocial and biopolitics studies, science and technology studies, arts performance studies, physical culture studies, sensory studies, animal relations studies and hybrid/hybridity studies.

Although originating in 'parent' social science and humanities disciplines, such as philosophy and sociology, health professional researchers have increasingly contributed to a new posthumanist theoretical orientation (e.g. Gagnon and Holmes 2016; Grant 2016; Holmes et al. 2007a, b, 2010; Holmes and Gastaldo 2004; Lapum et al. 2012a; Monteiro 2016; Roberts 2005; Sandelowski 1998, 1999, 2002, 2003). This orientation involves a number of underpinning understandings. As a whole, posthumanism challenges some of humanism's dogmas—its universal priorities and assumptions—that have persisted, arguably over centuries since the enlightenment

(Wolfe 2010). Posthumanism presents a critique of anthropocentrism—humans at the centre of the world as its most significant attribute; the prime holders of understanding and moral standing (Braidotti 2019a, b). In particular, a critique of the image anthropocentrism presents of human exceptionalism by destabilizing some of humanism's 'common sense' assumptions about the sovereign human subject (Castree et al. 2004a, b; Braidotti 2019a). It moves from a view of humans as autonomous, independent and considered actors to a much more relational view of them as dependent, networked and often automatic actors (Andrews and Duff 2019a, b). Specifically, under posthumanism, scholars realise that: (1) Humans have never been pure or independent and have always been 'more-than-human' (e.g. our bodies are not just 'us' and contain oxygen, carbon, hydrogen, nitrogen, calcium and other elements, our DNA stores data, and from the earliest pre-history, we have worn clothes and used simple tools). (2) Entities—including humans—cannot exist and be defined on their own, only in relation to one another (e.g. humans in relation to technologies and vice-versa). (3) Human-related phenomenon (e.g. wellbeing, health, skills, and numerous social or economic capacities) are not internal, personal or discrete qualities. They are 'modes of existence' that only emerge through expansive processes of relations between multiple human, biological, earth system, material and technological actors and forces. The result is that in posthumanist work scholars have torn down traditional conceptual divisions formerly established between the 'self', 'other' and 'object', recognising that human life is only one form of life and source of agency amongst many others of equal effectiveness in the world (Wolfe 2010). In short then posthumanism as a mode of thought is not preoccupied with the body escaping material bonds with a futurist—perhaps even science fiction—'posthuman' agenda. Rather, it shows the embodied embeddedness of the human in the material world. The body as an open system co-evolving with it (Wolfe 2010).

Posthumanism has in part been demanded by a wider posthuman condition emergent in society. As Mahon (2017) describes, this involves the unprecedented influence of technologies. For it is now with technologies that humans constantly change and develop as humankind, we being almost completely relational to them, with them defining our state or era. Braidotti (2019b) regards this as a fourth industrial revolution involving advanced technologies—such as robotics and artificial intelligence, nanotechnology, biotechnology—that blur boundaries between non-living material and biological things. She argues that the scale, speed and pervasiveness of this revolution will lead to planetary consequences—the sixth extinction. Hence, she argues that posthumanism approaches the heart of this problem and its building blocks and is the only way towards a sustainable and successful future. It is a navigation tool that helps us to understand the manifestation of advanced technological life and answer fundamental questions on its implications (Braidotti 2019b). As Andrews (2018) explains, with regard to health, changes have occurred in two realms in particular; (1) changing modes and forms of health consumption and (2) changing parameters of human life and health.

With regard to consumption, it is noted that through the workings of late capitalism, the commodity has completely colonized life in terms of spectacle - exploitation moving from production to consumption as well (Mahon 2017). Specifically, we live in a rapidly reconstituting world in which consumption has moved further to the foreground at an ever-accelerating pace (Andrews 2018). Indeed, human bodies move within an increasing range of environments in relation to an increasing range of objects with health consequences. Technological hardware (such as smartphones and tablets) and software (such as social media) give humans new forms of knowledge and awareness on health that bridge space and time, and become obsessions in themselves. New affective sensory pastimes entice and compel drawing human bodies into them, ranging from health activism to holistic and fitness lifestyles (e.g. Lea et al. 2015). Public health and health care organizations engineer root textures to their advice and care, adding multiple aesthetics and distractions that play to multiple senses and emotions (e.g. Solomon 2011). The proposition follows that humanism is not best placed to engage this faster moving sensory posthuman world of consumption, which is arranged and experienced differently and where the subject/self is under constant manipulation by affective forces. The stabilities humanism attributes to people, health and place no longer exist to the degree that they once did; and therefore, we need a new research lens that is able to bring this all into much sharper focus. As Braidotti (2019b) argues, posthumanism allows us to approach challenges of twenty-first century life in an informed way, including the burnout, fatigue, confusion, anguish, expectation, euphoria and joy all associated with this faster moving technologically mediated world.

With regard to human life and health, Duff (2014) suggests that there have been fundamental changes in recent years. First, under technological innovation (such as genetic engineering, bio-technology, nanotechnology, medical robotics, cell harvesting and cloning, pharmaceutical development and patenting and virtual reality) previously firm distinctions between phenomena (such as cells vs society, nature vs culture, natural vs artificial, medical vs economic) are increasingly breaking down, life itself becoming a 'natural resource' to be used and improved (Whatmore 2006). Duff posits that if human life is itself being reconceived, then research approaches also need to be also, and ones developed that redraw previous boundaries between phenomena, that are more aware of the many commonalities and relations between them, and more generally are better equipped to deal with these fusions of science, the body and society and the highly politically charged climate and debates that result (Whatmore 2006). As Braidotti (2019b) argues, posthumanism allows us to refocus on the human in posthuman times—to track the shifting ground of human and emergence of new understandings of human; the new relational and affective positons of humans. Though she readily admits that, for researchers and readers, this makes for excitement and anxiety in equal measures.

Rethinking Humanistic Concerns

Posthumanism clearly veers away from classical humanism, yet researching the social world whilst ignoring human traits, natures and needs, and human capacity for distinct, definitive and disruptive actions is not really possible, even for the most ardent or radical posthumanists (Castree et al. 2004a, b). Hence, another common understanding is that posthumanism is not exclusively 'post' (i.e. after) humanism, in that the term is an exaggeration and relations are maintained to earlier orientations (Castree et al. 2004a, b). As part of this, scholars do not throw away humanistic ideas (such as freedom, rationality, rights and individuality); rather, they are reconsidered beyond the academic constraints of humanism (Mahon 2017). In particular, pertinent to care, posthumanism tends to reimagine and extend popular humanistic concepts and concerns:

Identity, Meaning and Emotions

Elsie likes Broadview residential home and is proud to be a resident. For her, it not so much the facilities or even the care that matters, it's the sense of community she gets from joining in the activities there, that she her and her friends enjoyed in their early years. The once a month dances, theatre trips and regular film nights. They make the place for her. If you ask the staff, they will tell you a similar story. They enjoy creating different less institutional atmospheres, an achievement which makes them proud to be part of the organization.

In humanistic research, identity is typically presented as something consciously experienced and often reasoned in depth. In relation to health and illness specifically, modes of identity form around, for example, illness types and experiences, culture and belief systems and the production and expenditure of social capital. Posthuman thinking, however, adds the idea that identity is no longer only conceived in terms of self or externally attributed conscious association, and deep-seated complex verbalizable emotional ties and bonds. Instead, identities can be lighter yet no less powerful general 'feelings' of comfortable situatedness and attunement. Less-than–fully conscious associations attained through sensings in functional interactions and shared performative histories that are not always verbalized. Indeed, according to posthumanist thinking, the process of meaning-making—and perhaps ultimately even the formation of human values—might itself not solely be about narrative and known feelings, but may also be about physical processes and experiences that create conditions for the creation of meaning. As Duff (2014) suggests, senses of belonging (such as to groups, things and places) might be achieved, not only through conscious processes, but also through repeated physical and sensory exposures that become familiar and embedded. In the above fictional vignette, we see how for residents and staff alike, place identity is achieved at least in part around the sensory experience of activities, which are initially enjoyed and later valued.

Difference, Disadvantage and Oppression

Tamridge hospital goes out of its way to provide culturally-sensitive care. Many nurses and doctors come from the same East Indian population as dominates the hospital's catchment area. This means that staff are better able to understand and respond to clients' specific requests, as they relate to treatment but also aftercare, and conversations with family. The small things in the environment are also important, such as conversations between staff and clients in chosen languages, and with similar accents. The way that waiting rooms and communal areas have been decorated with particular cultural aesthetics.

Across humanistic health research, social difference, disadvantage and oppression (related to age, gender, ethnicity, class, affluence and socio-economic status) are understood as being consciously identity-based. Instead under posthumanistic thinking, they are known also to be based in physical performances of bodies expressing certain aesthetics, patterns and capabilities in relation to other bodies and objects expressing their own. These performative differences play into both unconscious and conscious repressive physical or verbal acts by the dominant group.

Some important insights into disadvantage and oppression have emerged from these understandings. Specifically, certain hierarchies that manifest in inequality might be a result of 'implicit bias' insofar as bodies are less in tune with bodies that interact with them less or do not easily fit their dominant patterns of interactions. This is possible because, as humans, we employ categories less-than-fully consciously as shortcuts (or perceptual habits) that involve familiarity/sameness and novelty just to process and understand the world we interact with (Andrews 2018). The aim of this kind of posthumanist thinking is absolutely not to evoke old determinisms or excuses for disadvantage and oppression, although it does present a way of thinking about society in ways other than it being constituted of constant, omnipresent overbearing exclusion and prejudice. In short, then, the aim is to realise the base relationality and potential of bodies as keys for achieving new understandings of, and creative means for, overcoming the negative consequences of hierarchical classificatory schemas (Andrews 2019). Moreover, to think about how, under the same principles, group affinities attained through shared affective senses of purpose and shared performances, might provide an effective response and a way to greater liberation (Haraway 1985, 1994). In the above fictional vignette of a local hospital, we see how familiarity (often subconsciously experienced) with other bodies—their looks and sounds—as well as with seemingly mundane materials, leads to a sense of affinity, ease and inclusion for staff and patients alike.

Agency

Marion, a ward manager tries to maximise person-centred care by allowing as much freedom of action as possible for patients in her care setting. This means supporting patients to do the little everyday things with personal possessions and to express their everyday behaviours as much as possible, no matter how seemingly small or trivial. The removal of rules, as much as is possible, whilst maintaining safety, allows for the continuation of natural repetitions that are part of all of our lives. Freedom to wear ones own clothes for example enables a form of freedom and expression as they say something about the person.

In humanistic health research, agency has always been bound up with conversations about the opinions, needs, principles, intentions, desires and options which underpin it; the commonality here being that these are consciously known and, for the most part, easily articulated. Nevertheless, under posthumanist thinking, the ontological basis of, and for, agency has been challenged. Posthumanists share a mistrust of the rational *cogito* and the premise of a self-aware, intentional subject, guided by clear objectives and capable of planned, intentional action in pursuit of these objectives (see Barad 2007; Latour 2005). It is not so much that scholars question the existence of such a subject, only that it should not always be regarded as the norm. They tend to downplay the importance of conscious processes while emphasising the role of affect, chance, encounters, events and relations in guiding action or agency in ways that are not always fully realised or articulated (see Braidotti 2013; Callus and Herbrechter 2012). Work in this vein often highlights the importance of *impulses* and *habits* in the life of (human) agents, whereby 'entrained' habitual responses to environmental cues that are mostly unconscious provide the motivation to act (see Bissell 2011; Dewsbury and Bissell 2015; Grosz 2013). This work has proven highly influential, shedding light on the habitual, less-than-fully reflective ways of acting and re-acting that frame health behaviours.

Of equal significance has been interest in the fundamental posthumanist contention that objects possess agency. Indeed, posthumanism radically opens the category of the actor to include humans as well as tools, plans, logics and processes as a means of accounting for the actions that objects exhibit (Latour 2005). A key theme has been the notion of what Latour (2005) and Bennett (2010), among others, call 'distributed agency'. This describes bodies and objects acting and experiencing together in networks. As influential as these arguments have been in recent health research, they however naturally open out to a host of unanswered questions about the causes or conditions of disparities in capacities, agencies or skills between actors in an assemblage. In sum, posthumanist analysis points to how not all forms of agency reach full signification, yet bodies and objects can communicate in other ways in the process of coming together, both aesthetically and affectively. Indeed, for the most part this 'coming together' involves singular and collective lines of action, spacing, intensity, speed and direction (McCormack 2013). In the above vignette we see the importance of clients' freedom of agency on the ward in PCC, and the agency of the seemingly mundane objects they might use and express with.

Narrative and Discourse

Jim talks to his clients, but sees the interaction as much more than information-sharing. He plays close attention to the way they tell their stories, and the way in which he responds, and the energy involved. For him, this is how 'real' conversations happen, when one is able to reverberate and express the energy of what happened.

Narratives about health, and in particular their constitution and meaning, are a key feature of an impressive range of qualitative research across disciplines. Narratives of course are regarded as expressive accounts of the actions and experiences of singular subjects, and it is this assumption that has most commonly featured in posthumanist critiques of narratives. These critiques typically point out that narrative/stories are always framed at least in part by earlier performances and experiences of physicality and sensing that are intersubjective, relational and often shared. Beyond this, in terms of its agency, narrative according to posthumanist thinking becomes less about content and meaning and more about context and expression: an act, event, happening or interaction (for example research on the practices and interactions in story-telling (Dean 2016)). In sum, narrative and discourse are no longer only about meaningful content. They very typically possess subject matter that is in some way performed 'out there' physically in life in the moment, and some cases can be partially and wholly about the immediate, sensory and performed aspects life. They are also bodily and spatial performances in themselves; of telling and conversing whereby rhythms, volumes, intonations, pitches and a range of environmental phenomena all interplay with meaning. We see this in the above vignette, where the professional and client's conversations are important physical events and actualities.

A Posthumanist Framework

In terms of the ongoing posthumanist project, modest recent attempts have been made, whilst still being mindful of its differences, to further extract common theoretical/processual understandings from posthumanist thought that might be useful in making sense of varied empirical contexts. Along these lines, Andrews and Duff have developed a posthumanist theoretical typology to describe the immediate more-than-human emergence and expression of particular aspects of life (Andrews 2019; Andrews and Duff 2019a, b). This typology articulates three spatial processes thought always to be simultaneously involved in any social reality: (1) life merging and expressed through relational material assemblages; (2) life enacted and performed affectively by open vital bodies with vibrant objects; and (3) life in immediate pre-personal, more-than-representational space-times. By the nature of its generality, this typology is outward facing, open and flexible, similar to what some posthumanist scholars term 'weak theory', which, as a loose description, is certainly quite apt. As Wright (2015, p. 292) notes:

> Weak theory sees things as open, entangled, connected and in flux. It is a practice of attending and attuning to things. Rather than closing down, categorizing, judging, modelling and getting things 'right', weak theory is open to possibilities, to surprises. It ponders connections and trajectories, and wonders what ways of knowing, of heeding and caring about things, are possible… weak theory supports partial understandings and multiplicity, and allows for both contradictions and inconsistency… weak theory promotes attention to affective assemblages, to the ways things, people, affects and places, with different trajectories, may come together, albeit in often tentative, inconclusive or evolving ways.

Here the framework is used to demonstrate potential rather than current record; what posthumanism might tell us about the processes always at play in the emergence and expression of PCC. Caring about personhood is evidently humanistic (about personal history, needs desires, dignity, experience, respect), yet any notion of PCC that proceeds from it has to be sensitive to the more-than-human make up of those meanings of care itself. Hence, a posthumanistic idea of 'person-centred assemblages of care' more adequately reflects the many materials, conscious and less-than-fully conscious bodies, practices, relations, events and affects necessarily involved in processes in situ—and the need to strike the right forms and balances between these.

Person-Centered Care Emerging and Expressed Within Material Social Assemblages

> *The proprietors of Marlyn House residential home make sure that residents rooms are personalized as much as possible, not only in terms of decoration but in terms of possessions, books, ornaments, places for handbags and favorite clothes. And more than this, they give their clients the opportunity to change them, rearrange them, handle them and talk about them as often as possible during each week*

As suggested, a fundamental position of posthumanism is its rejection of humanism's separation of the self and other. What emerges in its place is an 'object-orientated ontology'—or 'flat ontology'—which understands that human, biological and material entities are on the same level of existence (i.e. are of the same 'ontological type'), and co-evolve together. Further, it recognises that all social realities emerge through the generative capacities of 'assemblages' composed of these entities. According to scholars (see DeLanda 2016; Duff 2014), this working involves: (1) entities as the 'components' of assemblages that might stay in-situ, or move into or out of them; (2) the 'mechanisms' of assemblages (such as events and relations) that operate internally and within wider networks; (3) the internal 'processes' of assemblages (such as territorialisation, homogenization) continually in operation; and (4) the 'excessive' expressive vital outcomes of assemblages (such as affects) that emerge from them. The understanding is that new social realities are created through impermanent unities being reached within assemblages (in other words, through particular components possessing time-limited 'symbiosis' or 'sympathies'). Yet, as Müller (2015) suggests, all this happens without assemblages disclosing transcendental origins or all-embracing organising principles (see Deleuze and Guattari 1988). Overall, as a concept, assemblage has given posthumanist research a way to explain social realities that tells us a lot about their content, order, and how they are produced through the ways in which the world is arranged and bound. In the above vignette, we see the importance of residents rooms as personalised assemblages networked to the past and providing personalised anchoring and a sense of wellbeing in the present.

Personal-Centred Care Enacted and Performed Affectively by Vital Bodies and Vibrant Objects

The city-centre mental health unit organised a pet therapy session for their resident dog lovers; it involved a golden retriever, a Dashshund, German Shepherd and a Labradoodle. The dogs had different personalities and movements which suited particular clients. They ranged from being quite cuddly and loving to bright and energetic. The soft toys, balls, not to mention occasional treats, all had a role and were enjoyed by all

Body openness, sensing and potential are of central importance to posthumanist understandings. Scholars acknowledge how touch, hearing, sight, taste and smell lead to a powerful forms of knowledge achieved processually through the body and its hundreds of receptors and interactions. At the same time, however, as the aforementioned ideas on assemblages indicate, posthumanism thinks beyond the idea of the individual body as a receptacle, decentering it in favour of a broader capture of actors. Specifically, human bodies are recognised to change in step with, and through their relational encounters with, other biological and material actors (here matter not being considered as neutral or inert, but instead as possessing its own 'vibrancy'—capacities, agency and energy (Bennett 2010)). Overall then attention is paid to the interactive practices that vital human and other biological bodies partake in congress with vibrant material objects, the efforts they exert and the energies they create together. Deleuzian articulations of 'affect' have been key to understanding and articulating the nature of these efforts and energies (Deleuze and Guattari 1988). Indeed, Philo et al. (2015, p. 41) nicely summarise 'swathes of debate' on affect in human geography, commenting: 'affect is identified as a sensation which 'moves' between peoples and perhaps other life-forms, a flow of possibilities for feeling 'something' differently which circulates in the 'atmosphere' of a given place, preceding its localisation in an individual's emotional register (where it might also become available for reflection and 'wording') but possessing a definite generative potential (with the capacity to prompt *trans*-individual effects)'. Affects then are infectious transferences, and collective sharings, of 'feeling states' between bodies. Those affected move from one state to another; depending on the type affect, being either energising and strengthening or sapping and weakening. In the above vignette, we see these affects at play in a caring context, a therapeutic atmosphere emerging within the interactions of people, animals and things.

Person-Centered Care in Immediate, Pre-personal, More-Than-Representational Spacetimes

A dance class in dementia care might not seem an obvious choice but Karen the instructor facilitates easy to understand and gentle movements to songs of a period that the residents know well. Many become lost in swaying their arms to the rhythms, tapping to the beats, in following a partner.

Historically, social scientists have chosen to focus on the most explicit forms of conscious human acts and representations—perhaps because these are the most easily explained—frequently 'digging down' to theorise them and their consequences. As certain posthumanists argue, however, conscious human acts and representations constitute only a small part of life; most of what goes on in the world does not involve the rational cognition of words and images; instead, it is the constant, physical, automatic response to, and contribution to, material life (Thrift 2008). Indeed, as clearly evident in the previous discussion of affect, a posthumanist approach pays attention to the unspoken yet powerful immediacy, movement and practice of life, this particularly being the case for non-representational theory (Andrews 2018; Thrift 2008).

A research emphasis has hence developed on the base practices, flows and textures that create the pushes to and initial expressions of new life. At a micro-scale, this involves a focus on particular human and non-human movements, and qualities such as their lines of action, spacing and velocity (Andrews 2018). At meso and macro scales this involves a focus on how such movements, when dominant, create distinct 'bubbles' of space-time—or what some commentators describe as 'movement-space'—and the overall progression and becoming of these bubbles with particular velocities, rhythms and momentums (Merriman 2012). The challenge for scholars is thus how to convey forms of movement, expression, communication and inter-body solidarity that often do not involve full contemplation, signification or verbalization—i.e. 'non-representational' dimensions of life (Andrews 2018). In the above vignette, dance therapy provides a sense of joy in engagements with familiar sounds and motions in PCC.

Conclusion: Unpacking More-Than-Human Assemblages of PCC

As this framework suggests, professional health care, including all varieties of PCC, happens on a plane of continuous motion and experience. Whereas to care is typically regarded as a humanistic performance, the framework shows how care extends beyond humans interweaving different materials, relationalities, movements, distances, senses and emotions/affects (de La Bellacasa 2017). General attention is needed to understanding assemblages for PCC, and what is happening in them and their networks. This involves efforts to establish and describe causative pathways in the emergence of successful or unsuccessful PCC outcomes (such as those related to personal, family and community), often following particular interventions. A particular focus would be on the key components of assemblages that disclose in the course of these arising outcomes. Such attention would provide the most fundamental and relational view possible of professional 'expertise'/agency in action; this being realised in certain components, whilst supporting or countering other components. Further general questions, which involve forms of knowledge translation, include: Which types of assemblages provide the optimal PCC experiences and outcomes? Which types of political-economic, institutional, family and socio-cultural processes aid or obstruct assemblages for PCC? Are there optimal settings, family, community or practice assemblages, and what do they look like? How might social assemblages be (re)organized or augmented to create circumstances whereby clients, families, professionals and other partners are in fullest possession of their ability to act (hence which components and processes within a given assemblage could be utilized, which could be assisted, which could be countered, and which could be overlooked by health professionals)? What methodological innovations might help answer the aforementioned questions, and more generally unpack the processual aspects of assemblages, witness and animate their vital immediacy, and even intervene acting into and 'nudging' them in particular positive ways? Here, attention necessarily turns a little more introspectively to our own 'research assemblages', which are of course social assemblages in their own right, working in the same ways as any others but with their own specificities (see Fox and Alldred 2015). Comprised of their own bodies, tools, approaches and foci, and networked broadly to subject/social assemblages, each other and various institutions, these possess their own physical, affective and cognitive potentials. And if composed and arranged in specific ways, the potential to create positive outcomes both practically and politically (Fox and Alldred 2015).

References

Abdellah, F. G., Beland, I. L., Martin, A., & Mathney, R. V. (1960). *Patient-centered approaches to nursing*. New York: Macmillan.

Alharbi, T. S. J., Carlström, E., Ekman, I., Jarneborn, A., & Olsson, L. E. (2014). Experiences of person-centred care-patients' perceptions: Qualitative study. *BMC Nursing, 13*(1), 28–30.

Anderson, B., & McFarlane, C. (2011). Assemblage and geography. *Area, 43*(2), 124–127.

Andrews, G. J. (2018). *Non-representational theory & health: The health in life in space-time revealing*. Milton Park: Routledge.

Andrews, G. J. (2019). Health geographies II: The posthuman turn. *Progress in Human Geography, 43*(6), 1109–1119.

Andrews, G. J., & Duff, C. (2019a). Matter beginning to matter: On posthumanist understandings of the vital emergence of health. *Social Science & Medicine, 226*, 123–134.

Andrews, G. J., & Duff, C. (2019b). Understanding the vital emergence and expression of aging: How matter comes to matter in gerontology's posthumanist turn. *Journal of Aging Studies, 49*, 46–55.

Ash, J., & Simpson, P. (2016). Geography and post-phenomenology. *Progress in Human Geography, 40*(1), 48–66.

Ballard, C., & Aarsland, D. (2009). Person-centred care and care mapping in dementia. *The Lancet Neurology, 8*(4), 302–303.

Barad, K. (2007). *Meeting the universe halfway: Quantum physics and the entanglement of matter and meaning*. Durham: Duke University Press.

Bennett, J. (2010). *Vibrant matter: A political ecology of things*. Durham: Duke University Press.

Bissell, D. (2011). Thinking habits for uncertain subjects: Movement, stillness, susceptibility. *Environment and Planning A, 43*, 2649–2665.

Braidotti, R. (2013). *Metamorphoses: Towards a materialist theory of becoming*. New York: John Wiley & Sons.
Braidotti, R. (2019a). A theoretical framework for the critical posthumanities. *Theory, Culture & Society, 36*(6), 31–61.
Braidotti, R. (2019b). *Posthuman knowledge*. Cambridge: Polity Press.
Brännström, M., & Boman, K. (2014). Effects of person-centred and integrated chronic heart failure and palliative home care. PREFER: A randomized controlled study. *European Journal of Heart Failure, 16*(10), 1142–1151.
Britten, N., Moore, L., Lydahl, D., Naldemirci, O., Elam, M., & Wolf, A. (2017). Elaboration of the Gothenburg model of person-centred care. *Health Expectations, 20*(3), 407–418.
Broderick, M. C., & Coffey, A. (2013). Person-centred care in nursing documentation. *International Journal of Older People Nursing, 8*(4), 309–318.
Brooker, D. (2003). What is person-centred care in dementia? *Reviews in Clinical Gerontology, 13*(3), 215–222.
Callus, I., & Herbrechter, S. (2012). Introduction: Posthumanist subjectivities, or, coming after the subject. *Subjectivity, 5*, 241–264.
Carlström, E. D., & Ekman, I. (2012). Organisational culture and change: Implementing person-centred care. *Journal of Health Organization and Management, 26*(2), 175–191.
Castree, N., Coe, M., Ward, K., & Samers, M. (2004a). *Spaces of work: Global capitalism and geographies of labour*. Thousand Oaks: Sage.
Castree, N., Nash, C., Badmington, N., Braun, B., Murdoch, J., & Whatmore, S. (2004b). Mapping posthumanism: An exchange. *Environment and Planning A, 36*(8), 1341–1363.
Clissett, P., Porock, D., Harwood, R. H., & Gladman, J. R. (2013). The challenges of achieving person-centred care in acute hospitals: A qualitative study of people with dementia and their families. *International Journal of Nursing Studies, 50*(11), 1495–1503.
Cloninger, C. R. (2011). Person-centred integrative care. *Journal of Evaluation in Clinical Practice, 17*(2), 371–372.
Davis, S., Byers, S., & Walsh, F. (2008). Measuring person-centred care in a sub-acute health care setting. *Australian Health Review, 32*(3), 496–504.
de La Bellacasa, M. P. (2017). *Matters of care: Speculative ethics in more than human worlds*. Minneapolis: University of Minnesota Press.
Dean, J. (2016). Walking in their shoes: Utilizing go-along interviews to explore participant engagement with local space. In N. E. Fenton & J. Baxter (Eds.), *Practicing qualitative methods in health geographies*. Milton Park: Routledge.
DeLanda, M. (2016). *Assemblage theory*. Edinburgh: Edinburgh University Press.
Deleuze, G., & Guattari, F. (1988). *A thousand plateaus: Capitalism and schizophrenia*. London: Bloomsbury Publishing.
Dewar, B., & Nolan, M. (2013). Caring about caring: Developing a model to implement compassionate relationship centred care in an older people care setting. *International Journal of Nursing Studies, 50*(9), 1247–1258.
Dewsbury, J.-D., & Bissell, D. (2015). Habit geographies: The perilous zones in the life of the individual. *Cultural Geographies, 22*, 21–28.
Doherty, M., & Thompson, H. (2014). Enhancing person-centred care through the development of a therapeutic relationship. *British Journal of Community Nursing, 19*(10), 502–507.
Duff, C. (2014). *Assemblages of health: Deleuze's empiricism and the ethology of life*. New York: Springer.
Eaton, S., Roberts, S., & Turner, B. (2015). Delivering person centred care in long term conditions. *BMJ: British Medical Journal, 350*, h181.
Edvardsson, D., Winblad, B., & Sandman, P. O. (2008). Person-centred care of people with severe Alzheimer's disease: Current status and ways forward. *The Lancet Neurology, 7*(4), 362–367.
Ekman, I., Swedberg, K., Taft, C., Lindseth, A., Norberg, A., Brink, E., et al. (2011). Person-centered care—Ready for prime time. *European Journal of Cardiovascular Nursing, 10*(4), 248–251.

Finset, A. (2011). Research on person-centred clinical care. *Journal of Evaluation in Clinical Practice, 17*(2), 384–386.

Fors, A., Ekman, I., Taft, C., Björkelund, C., Frid, K., Larsson, M. E., et al. (2015). Person-centred care after acute coronary syndrome, from hospital to primary care—A randomised controlled trial. *International Journal of Cardiology, 187*, 693–699.

Fox, N. J., & Alldred, P. (2015). New materialist social inquiry: Designs, methods and the research-assemblage. *International Journal of Social Research Methodology, 18*(4), 399–414.

Gagnon, M., & Holmes, D. (2016). Body–drug assemblages: Theorizing the experience of side effects in the context of HIV treatment. *Nursing Philosophy, 17*(4), 250–261.

Grant, A. J. (2016). Storying the world: A posthumanist critique of phenomenological-humanist representational practices in mental health nurse qualitative inquiry. *Nursing Philosophy, 17*(4), 290–297.

Grosz, E. (2013). Habit today: Ravaisson, Bergson, Deleuze and us. *Body & Society, 19*(2–3), 217–239.

Haraway, D. (1994). A manifesto for cyborgs: Science, technology, and social feminism in the 1980s. In *The Norton anthology of literary criticism.*

Haraway, D. J. (1985). Cyborg manifesto: Science, technology, and social-feminist in the late 20th century. *Social Review, 80*, 65–108.

Hill, L., Roberts, G., Wildgoose, J., Perkins, R., & Hahn, S. (2010). Recovery and person-centred care in dementia: common purpose, common practice?. *Advances in psychiatric treatment, 16*(4), 288–298.

Holmes, D., & Gastaldo, D. (2004). Rhizomatic thought in nursing: An alternative path for the development of the discipline. *Nursing Philosophy, 5*(3), 258–267.

Holmes, D., Gastaldo, D., & Perron, A. (2007a). Paranoid investments in nursing: A schizoanalysis of the evidence-based discourse. *Nursing Philosophy, 8*(2), 85–91.

Holmes, D., O'byrne, P., & Coyte, P. C. (2007b). M3S squad: Considering cyborgs for future sexual health prevention programs. *Body, Space & Technology, 7*(1), 1–9.

Holmes, D., O'Byrne, P., & Murray, S. J. (2010). Faceless sex: Glory holes and sexual assemblages. *Nursing Philosophy, 11*(4), 250–259.

Jones, M. (2009). Phase space: Geography, relational thinking, and beyond. *Progress in Human Geography, 33*(4), 487–506.

Kitson, A., Marshall, A., Bassett, K., & Zeitz, K. (2013). What are the core elements of patient-centred care? A narrative review and synthesis of the literature from health policy, medicine and nursing. *Journal of Advanced Nursing, 69*(1), 4–15.

Kitwood, T. (1997). *Dementia reconsidered: The person comes first.* Buckingham: Open University Press.

Kontos, P. C. (2005). Embodied selfhood in Alzheimer's disease: Rethinking person-centred care. *Dementia, 4*(4), 553–570.

Kontos, P. C., Mitchell, G. J., Mistry, B., & Ballon, B. (2010). Using drama to improve person-centred dementia care. *International Journal of Older People Nursing, 5*(2), 159–168.

Kontos, P. C., & Naglie, G. (2007). Bridging theory and practice: Imagination, the body, and person-centred dementia care. *Dementia, 6*(4), 549–569.

Lapum, J., Fredericks, S., Beanlands, H., McCay, E., Schwind, J., & Romaniuk, D. (2012a). A cyborg ontology in health care: Traversing into the liminal space between technology and person-centred practice. *Nursing Philosophy, 13*(4), 276–288.

Latour, B. (2005). *Reassembling the social: An introduction to actor-network-theory.* Oxford: Oxford University Press.

Lea, J., Cadman, L., & Philo, C. (2015). Changing the habits of a lifetime? Mindfulness meditation and habitual geographies. *Cultural Geographies, 22*(1), 49–65.

Levine, M. E. (1967). This I believe…about patient-centered care. *Nursing Outlook, 15*(7), 53–55.

Mahon, P. (2017). *Posthumanism: A guide for the perplexed.* London: Bloomsbury.

Manley, K., Hills, V., & Marriot, S. (2011). Person-centred care: Principle of nursing practice D. *Nursing Standard, 25*(31), 35–37.

McCance, T., Gribben, B., McCormack, B., & Laird, E. A. (2013). Promoting person-centred practice within acute care: The impact of culture and context on a facilitated practice development programme. *International Practice Development Journal, 3*(1), 2.

McCarthy, B. (2006). Translating person-centred care: A case study of preceptor nurses and their teaching practices in acute care areas. *Journal of Clinical Nursing, 15*(5), 629–638.

McCormack, B. (2004). Person-centredness in gerontological nursing: An overview of the literature. *Journal of Clinical Nursing, 13*, 31–38.

McCormack, B., Dewing, J., Breslin, L., Coyne-Nevin, A., Kennedy, K., Manning, M., et al. (2010). Developing person-centred practice: Nursing outcomes arising from changes to the care environment in residential settings for older people. *International Journal of Older People Nursing, 5*(2), 93–107.

McCormack, B., Dewing, J., & McCance, T. (2011). Developing person-centred care: Addressing contextual challenges through practice development. *Online Journal of Issues in Nursing, 16*(2), 1–5.

McCormack, B., & McCance, T. V. (2006). Development of a framework for person-centred nursing. *Journal of Advanced Nursing, 56*(5), 472–479.

McCormack, D. P. (2013). *Refrains for moving bodies: Experience and experiment in affective spaces*. Durham: Duke University Press.

Merriman, P. (2012). Human geography without time-space. *Transactions of the Institute of British Geographers, 37*(1), 13–27.

Mitchell, G., & Agnelli, J. (2015). Person-centred care for people with dementia: Kitwood reconsidered. *Nursing Standard, 30*(7), 46.

Monteiro, A. P. (2016). Cyborgs, biotechnologies, and informatics in health care—New paradigms in nursing sciences. *Nursing Philosophy, 17*(1), 19–27.

Moore, L., Britten, N., Lydahl, D., Naldemirci, Ö., Elam, M., & Wolf, A. (2017). Barriers and facilitators to the implementation of person-centred care in different healthcare contexts. *Scandinavian Journal of Caring Sciences, 31*(4), 662–673.

Müller, M. (2015). Assemblages and actor-networks: Rethinking socio-material power, politics and space. *Geography Compass, 9*(1), 27–41.

Munthe, C., Sandman, L., & Cutas, D. (2012). Person centred care and shared decision making: Implications for ethics, public health and research. *Health Care Analysis, 20*(3), 231–249.

Murdoch, J. (1998). The spaces of actor-network theory. *Geoforum, 29*(4), 357–374.

Nolan, M. R., Davies, S., Brown, J., Keady, J., & Nolan, J. (2004). Beyond 'person-centred' care: A new vision for gerontological nursing. *Journal of Clinical Nursing, 13*, 45–53.

O'Rourke, D. J., Thompson, G. T., & McMillan, D. E. (2019). Ethical and moral considerations of (patient) centredness in nursing and healthcare: Navigating uncharted waters. *Nursing Inquiry, 26*, e12284.

Peplau, H. E. (1952a). *Interpersonal relations in nursing*. New York: Springer.

Peplau, H. E. (1952b). *Interpersonal relationships in nursing*. New York: G.P. Putman's Sons.

Parley, F. F. (2001). Person-centred outcomes: are outcomes improved where a person-centred care model is used?. *Journal of Learning Disabilities, 5*(4), 299–308.

Philo, C. (2007). A vitally human medical geography? Introducing Georges Canguilhem to geographers. *New Zealand Geographer, 63*(2), 82–96.

Philo, C., Cadman, L., & Lea, J. (2015). New energy geographies: A case study of yoga, meditation and healthfulness. *Journal of Medical Humanities, 36*(1), 35–46.

Pile, S. (2010). Emotions and affect in recent human geography. *Transactions of the Institute of British Geographers, 35*(1), 5–20.

Rankin, J. M. (2015). The rhetoric of patient and family centred care: an institutional ethnography into what actually happens. *Journal of Advanced Nursing, 71*(3), 526–534.

Roberts, M. (2005). Time, human being and mental health care: An introduction to Gilles Deleuze. *Nursing Philosophy, 6*(3), 161–173.

Rogers, C. (1951). *Client-centered therapy: Its current practice, implications and theory*. London: Constable.

Rogers, C. (1959). A theory of therapy, personality and interpersonal relationships as developed in the client-centered framework. In S. Koch (Ed.), *Psychology: A study of a science* (Formulations of the person and the social context) (Vol. 3). New York: McGraw Hill.
Rogers, C. R. (1961). *On becoming a person: A psychotherapists view of psychotherapy*. Boston: Houghton Mifflin.
Rokstad, A. M. M., Engedal, K., Edvardsson, D., & Selbæk, G. (2012). Psychometric evaluation of the Norwegian version of the Person-centred Care Assessment Tool. *International Journal of Nursing Practice, 18*(1), 99–105.
Ross, H., Tod, A. M., & Clarke, A. (2015). Understanding and achieving person-centred care: The nurse perspective. *Journal of Clinical Nursing, 24*(9-10), 1223–1233.
Sandelowski, M. (1998). Looking to care or caring to look? Technology and the rise of spectacular nursing. *Holistic Nursing Practice, 12*(4), 1–11.
Sandelowski, M. (1999). Troubling distinctions: A semiotics of the nursing/technology relationship. *Nursing Inquiry, 6*(3), 198–207.
Sandelowski, M. (2002). Visible humans, vanishing bodies, and virtual nursing: Complications of life, presence, place, and identity. *Advances in Nursing Science, 24*(3), 58–70.
Sandelowski, M. (2003). Taking things seriously: Studying the material culture of nursing. In J. Latimer (Ed.), *Advanced qualitative research for nursing* (pp. 185–210). Oxford: Blackwell.
Santana, M. J., Manalili, K., Jolley, R. J., Zelinsky, S., Quan, H., & Lu, M. (2018). How to practice person-centred care: A conceptual framework. *Health Expectations, 21*(2), 429–440.
Sharma, T., Bamford, M., & Dodman, D. (2015). Person-centred care: An overview of reviews. *Contemporary Nurse, 51*(2-3), 107–120.
Sharp, S., McAllister, M., & Broadbent, M. (2016). The vital blend of clinical competence and compassion: How patients experience person-centred care. *Contemporary Nurse, 52*(2-3), 300–312.
Sjögren, K., Lindkvist, M., Sandman, P. O., Zingmark, K., & Edvardsson, D. (2015). To what extent is the work environment of staff related to person-centred care? A cross-sectional study of residential aged care. *Journal of Clinical Nursing, 24*(9-10), 1310–1319.
Skaalvik, M. W., Normann, H. K., & Henriksen, N. (2010). Student experiences in learning person-centred care of patients with Alzheimer's disease as perceived by nursing students and supervising nurses. *Journal of Clinical Nursing, 19*(17–18), 2639–2648.
Solomon, H. (2011). Affective journeys: The emotional structuring of medical tourism in India. *Anthropology and Medicine, 18*(1), 105–118.
Thrift, N. (2008). *Non-representational theory: Space, politics, affect*. London: Routledge.
Travelbee, J. (1971). *Interpersonal aspects of nursing* (2nd ed.). Philadelphia: F.A. Davis.
Whatmore, S. (2006). Materialist returns: Practising cultural geography in and for a more-than-human world. *Cultural Geographies, 13*(4), 600–609.
Wolfe, C. (2010). *What is posthumanism?* Minneapolis: University of Minnesota Press.
Wright, S. (2015). More-than-human, emergent belongings: A weak theory approach. *Progress in Human Geography, 39*(4), 391–411.

Chapter 7
Towards a Research Agenda That Progresses Key Debates: Example II—Animating Emerging 'Skilling Space'

Gavin J. Andrews, Emma Rowland, and Elizabeth Peter

Abstract This chapter explores the potential for a more integrated and central geographical contribution to mainstream professional practice debates. As an example is describes how geography might play a fuller and more embedded role in understandings of 'skills'. After briefly reviewing traditional approaches to studying and understanding skill in health professional research, the chapter takes principles of non-representational theory and proposes some ways forward for future research through a deeper ontological understanding of skill and its processual, spatial emergence (summarised in the term 'skilling space'). Finally, at the end of this last chapter attention given to the broader future geographical research agenda on health care work.

Introduction

This chapter continues the theme of the last one to elevate geography beyond its current role as a 'useful' research lens and perspective by giving the discipline a more central voice in mainstream health professional practice debates. It showcases another area where and how geography might play a fuller, more embedded role that of debates on 'skills'. Specifically, after briefly reviewing traditional approaches to studying and understanding skill in health professional research, it takes principles of non-representational theory (NRT) and proposes some ways forward for future research through a deeper ontological understanding of skill and its processual, spatial emergence—summarised by us in the term 'skilling space'.

G. J. Andrews et al., *Place and Professional Practice*, Global Perspectives on Health Geography, https://doi.org/10.1007/978-3-030-64179-5_7

From Skills to Competencies: Traditional Perspectives on Levels and Capacities

Skills are discussed widely in health professional research due to the need for professionals to acquire a particular range of them—often through formal classroom education and on-site training—and deploy them in practice. Skills vary between jobs, grades and responsibilities, so any broad generalization of core skills required across health care is problematic and the literature consequently segmented. Certain areas of focused concentration, therefore, fall under particular occupations, including managers (Supic et al. 2010; Umiker and Umiker 2005), dentists and dental assistants (Miller and Scully 2015; Mossey et al. 1997; Robbins et al. 2006), radiographers (Castle 2006; Hardy and Snaith 2006; Lockwood 2016; Pratt and Adams 2003; Shanahan 2007) and nurses (Glen and Clark 1999; Jeffries et al. 2002; Perry et al. 2013; Smith et al. 2008). Other areas of focused concentration fall under clinical specialties that crosscut and incorporate various occupations including, for example, mental health (Delaney and Johnson 2006; Gijbels 1995), pediatrics (Gaies et al. 2007; Glover and Sussmane 2002; Oliver et al. 1991) and critical/intensive care (Gauntlett and Laws 2008; Reader et al. 2006, 2007). Finally, areas of concentration are also specific conditions and skills required to treat and care for them and their symptoms such as pain (Dalton 1989; Layman Young et al. 2006), dementia (Bourgeois et al. 2004; Chang and Lin 2005) and psychosis (Gijbels 1995; Mason et al. 2008).

Certain literature, meanwhile, deconstructs skills a little more as a category, distinguishing between ‘generic’ and ‘specific’ skills. Taking nursing as an occupational case in hand, with regard to the former, these are argued to include effective observation and communication (Bowles et al. 2001; Odell et al. 2009), critical thinking (Profetto-McGrath 2003; Wilgis and McConnell 2008), decision making capacity and execution (Shin 1998; Standing 2007) and physical movement, control, power and flexibility (Hignett 2003; Kneafsey and Haigh 2007; Wilson 2001). With regard to specific skills, these are argued to include documentation and the use of new technologies (Hobbs 2002; Jeffries 2001), medication knowledge and administration (Harris et al. 2014; Latter et al. 2000, 2001) and resuscitation (Davies and Gould 2000; Madden 2006).

Partly a response to errors in hospital care, shortfalls in hospital performance and noted capacity deficits of newly qualified professionals, all of which have become the focus of media attention, in recent years, ‘competency’—and its constituent ‘competencies’—has emerged as a key concept, somewhat surpassing skills in health professional research (Lima et al. 2014; Mid Staffordshire NHS Foundation Trust 2013). Competency is a little more expansive as an idea than skills. First because it implies a particular attained level of skill with some potential. Second, because it feeds into other central ideas such as safety and ethics, skill becoming part of a bigger picture. As the College of Nurses of Ontario (2014, p. 4) state:

> A competency is defined as the knowledge, skill, ability and judgment required for safe and ethical nursing practice.

Moreover, as King's College (2009, p. 1) state:

> Competency refers to specific capabilities such as leadership, which are made up of knowledge, attitudes and skills

As King's College (2009) describes, two main approaches in the literature have been used to conceptualize competence. (1) A behaviouristic approach has focused on tasks completed through practices that need to be observed (Dolan 2003; Manley and Garbett 2000; McMullan et al. 2003; Watson et al. 2002). This approach, however, has been critiqued for being reductionist, being focused on doing rather than knowing things; debate existing as to whether competence has to have been performed in order to have been demonstrated (i.e. potential to perform vs actual performance). (2) An holistic approach, meanwhile, has focused on clusters of connected knowledges, abilities and general attributes that are required to be effective and to carry/transfer competence into new situations. Hence, under this approach, a person's generic competence is understood to constitute more than the sum of their individual or site-specific competences (Dolan 2003; Manley and Garbett 2000; McMullan et al. 2003; Watson et al. 2002).

Competencies are often articulated as baseline requirements for various stages reached during professional education and training, such as at entry to first employment and at progression through the ranks (King's College 2009), and are aimed at, and tailored towards, specific clinical specialties (Jenkins and Calzone 2007; Mauk 2010; Polivka et al. 2008; Spielman et al. 2005; Van Leeuwen and Cusveller 2004). However, this said, research has managed to establish some general areas of importance. Zhang et al. (2001), for example, through empirical research established the ten competencies most often deployed in clinical events that had successful outcomes:

- Interpersonal understanding (a desire and capacity to understand others' predicaments, views and emotions—both spoken and unspoken)
- Commitment (being bound physically, emotionally and intellectually to a role and task)
- Information gathering (a desire and ability to seek to know more and to do better)
- Thoroughness (being careful and comprehensive)
- Persuasiveness (convincing and influencing in line with best practice)
- Compassion (sharing feelings, being concerned for feelings, and assisting accordingly)
- Comforting (supporting physically and psychologically to alleviate anxiety or pain)
- Critical thinking (making careful evaluations and judgements on causes, status and care)
- Self-control (remaining calm under stressful situations, controlling emotions and tolerating inconvenience)
- Responsiveness (acting promptly to inquires and needs—'being there')

In sum, and as Hird (1995) explains, the emergence of competency-based education has led to a number of trends, the most significant of these being an increase in workplace-based learning and simulation-environments, so that it can be demonstrated that what is taught in the classroom can be transferred to the 'real world'. Moreover, another trend being the expansion of professional standards for educational accreditation and professional registration that align with/require specific competencies.

Skills Bases and Skills Mixes: Traditional Perspectives on Skill Collectives

Closely tied to skills is the concept of 'skills base'; a geographical extension in that it describes the overall skills set possessed by groups as a resource within given locales. Whilst at a micro-scale, skills bases might be identified within institutions as available to their management; at a macro-scale, they might be identified within cities, regions and even countries as available to governments and other officialdoms to serve populations (e.g. Dornan et al. 2001; Godsell 2004; Jennett et al. 2000; Lehmann and Makhanya 2005). Skills bases are often discussed in the literature in the context of pressing needs or developments—for example, with regard to the emerging international trend of medical tourism and the safety and quality of local destination health care (Han and Hyun 2015; Turner 2010) and with regard to health professional migration and the consequences of the resulting 'brain drains' and 'brain gains' from and to specific locales and the rebalancing of human resources internationally (Astor et al. 2005; Buchan 2006; Kingma 2001).

Recent focus and debates have centred on the related idea of 'skills mix' (e.g. Crossan and Ferguson 2005; Jenkins-Clarke et al. 1998; Dubois and Singh 2009; Duffield et al. 2005; Harper et al. 2010; Jenkins-Brocklehurst and Tickle 2011; Lockwood 2016; Spilsbury and Meyer 2001). Here, assuming that particular occupations possess particular skills, research has often discussed proportions of different occupational categories needed within and between particular places (such as aids/assistants, nurses of various types and grades, technical staff, carers, doctors). Acknowledging, however, considerable skill/role overlap between different occupations, skills mix increasingly refers to combinations of skills and/or grades/seniority across occupations: macro surveys considering multiple sites (useful for system planning) and micro-scale surveys considering single sites (useful for improving bedside care) (Buchan and Dal Poz 2002). As Buchan and Dal Poz (2002) note, as twenty-first century health care becomes increasingly performance orientated, not to mention audited, finding the most effective skill mix becomes a priority and challenge for organizations and systems (often with specific goals in mind such as increasing workflow, patient outcomes or reducing costs). Particular confines and challenges in the current era are thought to include limited resources and availability, new regulations, established institutional traditions/cultures and sudden developments that require rapid reviews/changes in skill mix (including unforeseen staff shortages, escalating staff costs, quality enhancement directives, technological innovation and introduction, new procedures and programs, new regulations and health emergencies) (Buchan and Dal Poz 2002).

Acquiring and Deploying Skills: A Longstanding Theoretical Debate

The personal acquisition and deployment of skills has been subject to focused theoretical consideration and debate in health professional research. Certainly, the most high-profile contribution in this area has come from the leading nurse theorist Patricia Benner in the guise of her 'Novice to Expert' theory/model published in the mid-1980s—one of the most influential nursing theories of the twentieth century (Benner 1982, 1984). Justifying the need for her theory, Benner noted some contemporary phenomena emerging in healthcare at the time. On one level, she argued that because nursing is ever complex, and technology ever changing, it is no longer possible to standardise and routinise practice, yet still ever better and ever more skilled nurses are required in the workplace. On another level, she noted the over-reliance on theory and lecture theatres in nurse education of the time that, she argued, produces rule bound inflexible practitioners. These two things, she posited, demanded new understandings of how nurses learn and the different levels which they reach through time.

In terms of baseline ideas, Benner posited that nurses can obtain certain knowledge through classroom education (e.g. learning rules and principles) but importantly, can only become skilled/expert *at* practice through applying this knowledge, and gaining other practical knowledge, *in* practice over time (hence practice being a vessel to learning that has to be done to be mastered). Notably, a particular process was proposed by Benner; that applied experiences evoke critical thinking and change perceptions, resulting in increasing reality-based and intuitive decision making. Moreover, that applied experience results in decision-making that is 'holistic' in that it is momentary, often less-than-fully conscious taken yet draws on multiple knowledges. These propositions presented a challenge to the idea of 'rational reflection then action' and structured decision-making models more broadly, both common in nursing education and scholarship at the time (for overviews and critiques of these see Banning 2008; Lauri and Salanterä 1998; Thompson 1999).

In terms of detail, Benner's model maps directly onto the 'Dreyfus Model of Kill Acquisition' (Dreyfus and Dreyfus 1980)—originating from the discipline/field of operations and educational research—and describes five levels that most nurses gradually reach. As they move across these five levels, fed by 'real world' experiences, nurses are thought to demonstrate decreasing dependence on abstract ideas, increasing levels of critical thinking and increased use of intuition:

Novice: At this initial stage, nurses have no previous experience of the situations in which they perform. They do not exercise—and are not allowed to exercise—personal judgements in clinical matters. Tasks are attributed to them. They often lack confidence to practice, and must adhere to cues, rules and plans. They have to focus and concentrate on their mentors and on all aspects of their job.

Advanced beginner: At this second stage, nurses display marginally acceptable performance, which is based on their past experiences. They might be skillful in some areas but still need occasional support and cues. They have limited situational perception, and they do not typically prioritise or join up particular aspects of their work.

Competent: At this midway stage, nurses demonstrate confidence and efficient behaviours. They are able to cope with multiple activities and increases in workload, and they can act in relation to long term goals that they understand. They can plan and create routines based on analytical contemplation of a situation. Most often, they do not require supporting cues.

Proficient: At this penultimate stage, nurses possess holistic views of situations they encounter, as well as of long terms goals (thinking about all the reasons and consequences), which quickens and improves their decision making. They can recognise non-expected outcomes and respond accordingly. They can recognise relative importance, prioritise, deviate from plans and adapt.

Expert: At this final stage—ideally a career-long 'steady state'—nurses have a deep intuitive understanding of total situations and can centre in on the relevant region of a problem without having to consider all possibilities. They do not rely on rules and guidance and can perform fluidly, flexibly and proficiently. They possess a highly skilled analytical ability even in the case of situations they have never previously encountered. They also possess visions for future possibilities.

Popular in nurse education and research, the Benner's model has been described, applied and extended over three decades (see, for example, Courtney et al. 2008; Gatley 1992; Haag-Heitman 1999), as well as being revisited by Benner herself and tested with empirical data (Benner 2004). However, despite its influence, it has certainly not been uncritically or universally accepted. In a gentle critique, Gardner (2012), for example, notes three ironies in Benner's work. The first is that, even though it is one of the most influential nursing theories, it is surprisingly light on theory and is very descriptive (Gardner noting that Benner herself suggested, for example, that nurses should not look too deeply into the 'mysteries of practice'). The second irony is that Benner's model that she is famous for is not her own thinking, and it is a minimally extended empirical application of Dreyfus and Dreyfus (1980). The final irony noted by Gardner is that Benner outlines a model, but, at the same time, proposes that models in general are of limited value in education—only to novices—and that 'practice rules'.

A more aggressive and theoretically based critique of Benner's work has come from English (1993), who, as a positivistic researcher, takes issue with Benner's phenomenological leanings. On one level, English is not convinced by Benner's handling of the idea of 'expert', which he thinks is rather vaguely conveyed in her work. He notes some important unanswered questions, for example: When does one become an expert exactly? What evidence suggests someone is an expert (or at any of Benner's five stages)? Are some experts better than others? Indeed, English considers Benner's mapping of particular years of learning onto particular stages to be crude and inaccurate. On another level English contests Benner's ideas on how expertise works, in particular challenging her notion that it works predominantly from an intuitive base (this he claims does not always happen and is a foolhardy claim in the current era of evidence-based practice). Indeed, English observes that nurses do not 'recognise' situations and respond reflexively; rather, they more often consciously 'identify' situations and act accordingly (their repeated exposure to incidents they consciously resolve producing clinical knowledge). In other words, they have a pretty accurate internal picture of what something should look like, a picture to which they refer before they respond. Adding to this debate, other scholars have noted the considerable theoretical separation of Benner and English, discussing the need for any model of learning to have, as English would want it, objectivity, generalizability and predictive power (Darbyshire 1994; Paley 1996).

Efforts have been made to move beyond support and criticism of Benner's novice to expert theory by coming up with fundamental modifications and/or alternatives. Indeed, Thompson (1999) suggests that the debate whether systematic cognitive reasoning/processing or humanistic intuitive practice is responsible for skill acquisition and deployment is largely redundant, because rather than these being separate/alternatives, they likely occupy positions at the opposite ends of one continuum. Recognising this, Gobet and Chassy (2008) try and find a mid-way between Benner and English by proposing a new theory of 'expert intuition' where intuitive perception/reaction and conscious problem solving are both recognised and accounted for and, moreover, are recognised to be interrelated. They argue that intuition does not 'just happen' (holistically), as Benner describes, but is far more complex. Their idea is that health professionals learn patterns or chunks of information that, stored in their long-term memory, they use as templates for comparison that code new information.

Towards Emergent 'Skilling Space'

Although various adjustments and alternatives to Benner's ideas have been forwarded over the years, it seems like these are, for the most part, quite 'static' in that they miss explanations of the active process of skill acquisition and deployment. Indeed, it is at the theoretical intersection of contemplation and intuition that we intervene, proposing that perspectives and understandings from the contemporary geographical tradition of non-representational theory (NRT) (as introduced in Chap. 2) might prove insightful. As suggested in Chap. 2, NRT is an approach that draws variously on Deleuzian metaphysics, new materialities, sensory and relational social science and other paradigms, to focus on the 'bare bones' practice of life. It animates what is happening processually and physically in the creations of space-time, often evoking forms of awareness and knowledge that do not involve full contemplation or verbalization (Andrews 2018; Thrift 2008). NRT, we argue, helps develop a deeper ontological understanding of skill and its immediate spatial emergence in health care.

Although skill acquisition and deployment is not a common subject of scholarship in NRT, work by Patchett (2016) and Ingold (2000) on craft skills has been key in this area. At one level these authors' ideas are consistent with Benner's. For them, successful skill acquisition and deployment is a practical matter of sameness and replication. Skill is a practice, so it can only be acquired through practice. Most notably, it is not itself something written, so it cannot be acquired through reading and writing (or any other such representations). However, like Gobet and Chassy, these authors conflict with Benner's idea that skill is 'innate', they arguing instead that skill is 'grown'. Where they move beyond Gobet and Chassy's critique, is by progressing a spatial argument: Skill acquisition and deployment necessarily involves the body in an environment—specifically, embodied capacities and embodied awareness of, and interactions with, environmentally situated agents. In other words, skill acquisition and deployment is an environmentally emergent relational process through interactions within particular assemblages of bodies and objects. Most recently, providing further elaboration, Patchett and Mann (2018) expose of 'five advantages of skill' in their anchor paper to a special issue on 'geographies of skilled practice' published in the journal *Cultural Geographies:*

1. *Skill is practical*: Skill is practical in that it helps do something (a task) and achieve something (an endpoint). The authors mention, in particular, that through repetition, skill might become habitual, making it more easily applied and replicated (this challenging the notion of human habit as unthinking and mechanistic). They also argue that rather than thinking of habitually practiced skill as thought processes retreating and bodily responses emerging, it might be thought of as 'a unitary circuit of body-brain-environment' involving 'finely tuned and highly creative forms of thinking-acting-responding' (2018, p. 24).
2. *Skill is processual*: Rather than working through consecutive discrete phases with constant reference to others and rules, people often work emergently and responsively in a continuous variation (with others and rules). The authors suggesting—in line with Ingold (2013)—that this is why skilled practice is better followed than predicted, and why practice adapts and mutates over time.
3. *Skill is technical*: Rather than being discretely operational in single bodies, skill is technical and distributed across bodies and tools that work and evolve together. Tools assist but also act as centres for particular body know-how and action. The authors note how, rather than leading to the loss of humanness and skill, tools encourage new forms and levels of skill (a new embodied intelligence).
4. *Skill is ecological*: Skill is ecological because its practice encompasses entities and events beyond the single body; multiple evolving and environmentally situated relations between bodies, minds, objects and rules. Skills being, for example, grown and incorporated—oftentimes following an irregular and precarious path—rather than being innate or acquired.

5. *Skill is political*: Skill is political, largely realised in the relationship and flow between scales: the micro where it is emergent and practiced, often less than fully consciously expressed, and the macro where it is consciously discussed, regulated, enforced and assessed. Often the agency of the former being relational to—positively and negatively—the texts and power of the latter.

Providing additional insight and explanation into what academics might look for in skill acquisition and deployment, Ingold (2013) talks about the need to move from the mainstream idea of 'making through thinking' to the idea of 'thinking through making'. Ingold argues that the former is the most typical way we understand how things are produced. Theory leads, then practice follows as people project and impose ideas on objects (an artifact hence being a materialization of a prior thought). Ingold argues that this idea—of only looking backwards to an idea, of making being a projection of a ready-made thought—does not adequately account for human creativity and innovation. Instead 'thinking through making' is based on the idea that process is itself creative; process generating new ideas and things. On one level during the flows and transformations of materials. On another level during the movement of the imagination under sensory awareness. Or as Ingold puts it, "making is an ongoing binding together of natural flows and sensory awareness, what we see being an outcome of that performance" (p. 5). Ingold hence argues that academically we thus need to 'think creating forwards', rather than backwards. By this he means we should not always attempt to (re)discover or back-check the original idea. Instead, we need to join with the movements of making; the movements of materials and awareness as they are feeling their way ahead in space and time.

Using these ideas, in the remainder of this chapter, we want to 'think creating forwards' with regard how skill emerges in space and time, or in what some scholars of NRT think of as 'whole onflow'; the emerging frontier of existence, the leading edge of space-time (Andrews 2020). The understanding is that there exists an unbroken unified stream of physical and experiential becoming. For an infinitely small moment, all life is brand new, physical and less-than-fully-consciously acted an experienced. It is this infinitely small moment that humans occupy (and no other): a moment happening all the time and everywhere. Whole onflow makes the world subject to immediate change and is where human spirit and innovation is exercised. The question arises as to which particular qualities of whole onflow might be part of the skill acquisition and deployment process? Using Andrews' (2018) typology, we suggest the following eight, each section beginning with an illustrative fictional vignette. Whilst all eight are physical, engaging and felt, some are more part of the forwards movement of onflow (speed, rhythm, momentum), some are more like underlying orientations and contributions (vitality, infectiousness), whilst others are more about occurrences (imminence, encounter, stillness).

Speed

John, an emergency responder, always moves quickly in response to a call. An extra minute here and there can make the difference between life and death. He and his crew do not rush however. They stay in control because their safety and the safety of the public is of paramount importance. Once John arrives he must quickly acquire information, assess the situation and start to act. Over the years John has gained a level of seniority and uses his experience to mentor novice responders. He coaches them through processes so that in future they can undertake them independently, whilst demonstrating technical safely and promptness. He will often ask them to observe him and talk them through his thinking, eventually letting them take over from him whilst giving them further direction and maintaining their critical mode of thinking. With occasional verbal nudges he coaxes them along the correct paths "yes, vital signs first and straight away" "okay, now look at his arm" "right we have got that, now lets move to his lower body". Eventually, with experience their own practice speeds up and improves.

Speed is understood scientifically as the rate of motion of an object measured in terms of how much space is covered by it in a specific time (such as meters per second). Any object possessing mass will necessarily have to accelerate to reach a specific speed. Once that object is in motion, its speed might be stable (constant), increasing (accelerating) or decreasing (decelerating) over a given duration. Human subjectivity becomes involved when the speed of an object is compared—often visually—to the speed of other objects; this relationality provides descriptors/attributes (such as something judged to be moving 'fast' and 'slow'), the same being the case for the speed of events (such as something judged to be occurring 'quickly' or 'slowly'). Indeed, for humans, these attributes can also be a practice (going fast or slow or going faster or slower), an aim (to go fast or slow, or to go faster or slower) or a sensation (feeling fast or slow, or feeling faster or slower). On a collective level, many social phenomena possess complex relationalities and subjectivities in terms of their speed (such as market development, policy implementation, urban development, manufacturing output, consumer purchasing). In practice, however, because life necessarily has movement, all life possesses some speed.

As the fictional vignette above shows, the acquisition and deployment of skill involves speed. On one level the relative speed of bodies and objects involved in its emergence. A student might experience the speed of activity in their practice, the speed of their own work in relation to others. They might seek to moderate their own speed and occasionally the speed of their mentors. On another level is the speed of the overall process. Students might be aware of the speed of their own learning, the learning of their fellow students, and the speed at which they are progressing through their training. Speed is critical to educational bodies and health care organizations who need to work to certain timeframes, yet the success of skill acquisition or deployment is dependent on the correct speed being attained.

Rhythm

Sally, a mental health nurse, used to work in an institutional setting for mental heath care. She liked it and became familiar with the rhythms of the working day—breakfast, rounds, admissions and interventions—but after a few years she got bored with these routines which failed to engage her. She how works in community settings visiting schools. On one level she now experiences new daily rhythms in travel between locations, going from one school to another, on to a meeting in a café and so on. A crisis however, such as a lost child or even a suicide attempt, might disrupt this daily rhythm and create another, often more frantic and focused. On another level are the rhythms in Sally's conversational practice; the therapies and counselling she provides for youth. These rhythms enroll them, keep them involved and help them. Sally is now a mentor to other community mental health nurses and she has taken a few along on her visits, showing them how to move from one question to another: "how do you feel" "why do you feel that way" "when it happens again try to think of a better outcome" "do it with me now". For Sally its all about being mindful of where clients need to be in the session; responding to their needs therapeutically in the flow of their conversation. In this way rhythm helps her do as much as she can in the limited time available to her.

Rhythm is understood scientifically as a regularly patterned movement identifiable by its timing/spacing; repeated intervals of either alternating strong and weak periods or alternating progressions and stops. These intervals create a pace to the rhythm. Resembling this scientific understanding of rhythm, but not as mathematically precise, are numerous human expressions and experiences of rhythm. Spatially bodies might experience and participate in the rhythms of homes, workplaces, caring spaces, consumer spaces, neighbourhoods, cities and regions and temporally the rhythms of moments, hours, days weeks and years. Humans participate in and experience rhythm through being immersed in it to some degree physically and mentally. Rhythm is sensed, registered, internalised, learned and acted often less-than-fully consciously, the body becoming in-tune with it. Its regularity provides a comfort and incentive to move along with other bodies and objects, it feeling 'natural' to do so (or the opposite being the case when it feels awkward, difficult and unnatural to be 'out of sync' with the dominant surrounding rhythm).

Participation in rhythm is far from trivial. Without the body's ability to act less-than-fully consciously with and within rhythms, every single aspect of everyday life would require pause and constantly consideration. On a collective level less-than-fully conscious involvement in rhythms is a fundamental way in which humans inhabit and produce space. Lefebvre (1992), for example, talks about how rhythm creates an 'animated space' amplifying the space of bodies involved, making them larger in terms of apprehension, particularly in relation to those observing. Moreover, for Bergson (1911), the synchronization or conflict of different co-existing social rhythms might lead to different degrees of mutual relaxation or tension amongst bodies on different levels of consciousness. Ultimately then, as Deleuze and Guattari (1988) posit, rhythmic relations and expressions between social milieus create 'territories' of different style that are not hard barriers but motifs; counterpoints to each other.

As the fictional vignette above illustrates, clinical environments possess and work to rhythms that are critical to skill acquisition and deployment individually and collectively, whether these environments be community or institutionally based, and whether these rhythms be over minutes, hours, days, weeks or years. Meanwhile counterproductive rhythms might disrupt such processes. A health professional learning might experience different rhythms in different places, of different tasks and at different times throughout a working day. They might feel themselves gradually becoming more in-sync as they learn and are able to fully participate. Ultimately, they might experience the longer-term rhythm of their overall development.

Momentum

Amy is a senior nurse on a busy general hospital ward caring for a wide-range of clients. She finds the working day a challenge particularly the latter half which can be taxing and tiring. Some days are better than others. When the team is working well together, it has the effect of picking her up and carrying her along. When the team is not functioning properly, when one problem cascades into another, the day can seem very long. Amy has a trick or two she shares with her fellow nurses and doctors. Its using simple visualization techniques when injecting clients who suffer from needle phobia (perhaps an IV line for medications or flushes). Important to her practice is keeping the visualization going with simple questions so that clients do not enter conscious contemplation of their situation and focus on their fear or what is happening. She asks "where did you go on vacation last?" "oh Cuba" what was the beach like?" describe it to me". All the while Amy maintains procedural momentum with the injecting task in hand.

According to the laws of mechanical physics, linear momentum is a motion force that is a combination of an object's mass and its velocity ($p = mv$). In terms of practical application, engineering momentum is most often considered in terms of the energy required to create it (to get objects moving forwards) and to kill it (to stop objects). Momentum, however, also occurs in the social world albeit, as in the case of speed and rhythm, in forms that are less mathematically precise and more subjective. Three forms in particular exist. The first, behavioural momentum, describes the tendency for human behaviour to persist due to its own physical and mental impulses and/or in how one human activity potentially leads onto another with implications (such as in the contexts of drug use and career trajectories). The second form, circumstantial momentum, is where due to the imposition of structures or changing personal or social circumstances, one change for humans can cascade into further changes, sometimes in positive directions but oftentimes in negative ones (such as in the context of losing a job). The third form, felt momentum, transpires through, and is experienced as, the basic physicality and force of bodies and objects moving in particular directions. Bodies can experience, for example, the agency and feeling of being swept up, of moving with and adding to the prevailing momentum (and thus the sheer joy of momentum itself, or discomfort of momentum if the direction is unwanted). Alternatively, one can experience the agency and feeling of inertia (there being a lack of momentum), or of moving and working against the prevailing momentum. These positive and negative experiences are often in relation to what is going on immediately, although they can also be about longer term events (such as feeling one is 'stuck' or 'in a mire').

An important category of behavioral and felt momentums is 'performance momentum' (in lay terms know as being 'on a roll'). Performance momentum is a dynamic intensity characterised by an enhanced state of motion, grace and success. It involves the synchronisation of mental sub-systems (such as excitement and confidence) but also the positive interchange and recycling of emotional and physical energy; the force and intensity of performance momentum propels individuals and groups towards their goals, allowing them to perform to their maximum potential (Adler and Adler 1978). Meanwhile, one residual outcome of performance momentum can be a positive psychological mindset and physical disposition; an affective memory that remains in individuals or groups after the end of one event until the start of the next. This facilitates the reintroduction of performance momentum once it commences, allowing 'the hot hand' to continue.

As animated in the previous vignette, skill acquisition and delivery itself has some momentum (like it might equally have inertia), highly interrelated circumstantial, behavioral and felt and performance momentums carrying a trainee forwards in their acquisition and deployment of skill. Clinical environments of various forms possess momentum as a whole and there can be procedural momentum within and between specific tasks. Through momentum skill carries itself forward, ultimately momentum being a factor in whether skill has positive effects and the extent of these effects. Conversely, skill meanwhile might equally stop an unwanted or destructive momentum in a clinical environment and help replace it with a new direction. More generally, there is momentum in the overall educational experience, and ultimately momentums in the lives and development of health care institutions themselves.

Vitality

Sean is a cardiac rehabilitation specialist who although hospital based, has the opportunity to organise outdoor fitness activities for his clients in a nearby park. This environment, with fresh air and shade and smells of trees, along with occasional sounds and sights of passing wildlife, provides an experience of care that is far more holistic than the activities alone. Sean has taken this experience and transferred what he can of it to the hospital gym; vibrant colours, relaxing sounds—which certainly make sitting on a static bike a far better experience than it once was. Sean realises that his clients' fitness is crucial to enhancing their autonomy. He pushes them to maximise their own vitality but not to the extent that the process itself becomes dangerous. On a psychological level its all about giving them hope through adjusting how they see themselves in relation to their environments.

A scientific understanding of vitality stresses the diversity and strength of a group (for example, facets in a biological population). In NRT, however, vitality refers to something more. Here, it is about the aliveness and buoyancy of the living and non-living worlds and their collective potential; that there is something about them that is not explainable by mechanical and chemical physical forces or laws. With regard to the living, scholars draw on vitalist philosophy, which recognises the exceptional qualities possessed by all living things: their essential spirit, spark and energy (Philo 2007). With regard to the non-living, direction is gained from new materialist thinking, particularly from ideas on 'vibrant matter' (Bennett 2010). This describes the capacity of things—from core materials to more complex objects—to act as quasi-agents with their own tendencies, trajectories and forces. Both of these lines of thinking recognize that when encounters happen within and between the living and non-living, an energetic animation results that gives life a range of qualities including its richness and diversity, its endless capacity to move and evolve on its own impulses, yet also its instability and unpredictability. As an intellectual idea, on the one hand, vitality provides a relief from deterministic and mechanistic positivistic academic thinking by providing a reevaluation of how to understand things—not as definable by their 'properties' but by what they 'do' as more than the sum of their parts in energetic flux within the milieus within which they exist (Philo 2007). On the other hand, vitality provides relief from social constructionist academic thinking, obsessed with meaning and identity, by shifting attention to the fundamental make-up, character and push of life.

The last vignette reflects the reality that vitality is critical in how skill is acquired and deployed. Vital learning bodies, vital client mentors and workers bodies and vibrant clinical objects—from the basic to the complex—all working energetically together (often including vital or vibrant viruses, bacteria and drugs of various descriptions). Together they produce and outcome. This is an excessive dimension of skill; one that is powerful, yet can not be fully understood and accounted for.

Infectiousness

Dan works in a pediatric ward in a local hospital. It is one of the leading wards of its kind in the country and the staff are well aware of what they have together created. In the business of the day they work energetically and enthusiastically and efficiently feeding off each other's cues and boosting the overall effort. Now a mentor, Dan tells new staff about the kind of environment they will be working in. His enthusiasm is palpable and obvious, setting them off on the right positive note. Dan did some qualitative research on his workplace. In the interviews he conducted his colleagues had difficulty putting their everyday experiences into words. Some talked about 'metaphysical energy', some talked about 'positive vibe', some talked about 'good emotional tone', some talked about 'physical optimism'. Others meanwhile mentioned contrasts they encountered in other places such as 'inertia', 'indifferent sluggishness'. Despite these differences in words, all knew however exactly what they were referring to, and all were talking about exactly the same thing.

A scientific understanding of infectiousness typically centres around the transmission of disease via microorganisms. Infectiousness can, however, occur and be experienced in the social world as something that involves and enrolls humans. In NRT, infectiousness of human action and feeling is understood specifically through a Deleuzian reading of 'affect' (Deleuze and Parnet 2006) which involves a radical rethinking of the nature of the human body. Deleuzian philosophy asks not what the body is (such as how it is composed and works) but what it can do (its potential) and what it does (the results of its expenditures of energy), both individually and 'transpersonally' (produced through, circulated amongst and shared between bodies) (Duff 2014). As Duff suggests, on one level, the body's facets—such as expressions, habits, movement forms, ways of speaking—are folded into the body as transmitters. On another level, the body's numerous sensory detectors are open to detecting the facets of other bodies. From this perspective then, the body is permanently amiable to relations with its surroundings, both signalling and open to signals.

With this idea of the body as a starting point - and extending earlier discussion of affect in this book - affect then can be understood as a transitioning of the body and the process whereby it is affected by other bodies, modifies and affects further bodies (see Thrift 2008). The transition from one experiential state of the body to another encompasses changes in its energy, which is either amplified (positive affection or a joy/laetitia in the body) or drained (negative affection or sadness/tristitia in the body) (Pile 2010). Importantly affects are not fully known to or reasoned by individuals as they occur, but are experienced less-than-fully consciously by them, revealing on a somatic register as a change in their 'feeling state'. Affect, then, is a 'passion' of both the body and the mind experienced powerfully if only vaguely realised.

Environmental contexts are critical to the production and experience of affect. As Duff (2014) suggests, in Deleuzian terms, 'milieus' assembled from diverse human other living and non-human elements help generate affects in particular territories. Affective environments are hence transmitted—being soaked up and registered by humans both singularly and collectively—and experienced as prevailing 'atmospheres' (Anderson 2009). Notably, affect might also bridge scales; it being possible, for example, for affects to be generated and felt between thousands of bodies—often through technological mediation and media that transports them over physical distance—at global levels.

Importantly, as Hall and Wilton (2017) suggest, what drives human bodies into affective relations with other human bodies, and more broadly into affective environments, is a basic 'desire' (cupiditas)—that is natural but not always consciously realized—to 'do'. To move beyond the single body and be part of bigger happenings, the bigger physical picture of the world. Offering opportunities to fulfill this desire, affective atmospheres are thus alluring nodes of basic feeling that invite humans in—and once in, to respond (Anderson 2009). Notably, armed with lay versions of these knowledges, state and commercial interests often purposefully create and re-produce affective environments with specific atmospheres, providing familiar textures to peoples' lives often through, for example, attention to the sensory attributes in standardized forms (Thrift 2008).

The vignette above illustrates how affect assists skill acquisition and deployment through producing something that is more than the sum of its parts. Bodies working together, working off each other, together creating a learning or healing energy. This feeling of joyful physical investment, of being part of a bigger picture, also enrolls further investment, and thus increases success and productivity. Meanwhile, through the purposeful designs and layouts of clinical environments—which are increasingly open plan—health care organizations seek to create a positive working atmosphere.

Imminence

Mary works in an emergency room. It's a stressful place because she and others never know how much and what type of care they will have to provide next, and what the outcome might be. To be successful, more than anything she needs to be prepared. She knows something is coming, sometimes expected, other times unexpected. That's the nature of her job. Mary's mentor was key in getting her ready for the unknown, after key events debriefing and reviewing with her what went on. This occurred only last week following a traffic accident. "Did you feel you had the space prepared?", she said, "what will you do next time?". Through this process Mary feels better ready, more confident in her workplace and role; she can imagine many occurrences and what she will do.

Imminence refers to the close future arrival of something. In terms of likelihood, that arrival is certain/impending, or at least thought to be. What exactly might be imminent in life varies greatly, from the initially non-physical (such as the pre-dated arrival of laws, rules or policies) to the physical (bodies and objects). Imminence is particularly important in terms of how the arrival is expected or unexpected; how bodies and objects are prepared or ill-prepared for it, and the actions they take or do not take prior to and on arrival.

In NRT, imminence is important as a state and experience in itself. Imminence occurs because other qualities of whole onflow (such as rhythm, momentum) possess energy and forwards movement thus give some notice. Indeed, NRT thinks of the world as being in a state of permanent imminence, where everywhere in it is forever incomplete, and everywhere in it is open to something new, that will always arrive. Moreover, as Andrews (2018) posits, NRT strives to think of life being configured as a series of infinite 'ands', which add to the state of affairs. Thus, NRT is concerned with what might be expected next, and after that; being just as concerned with the jumps forward as the events themselves (the spacing that enables and introduces the next moment).

This all goes back to the way in which NRT conceives and understands the body. As Andrews (2018) suggests, rather than thinking about the body as something that 'is' (with particular qualities), it thinks of it as something that possesses potential (that might do)—a body literally waiting for something to happen and waiting to react in some way. Moreover, a body that feels and remembers; affective memory occurring as a momentary virtual event that helps bodies 'prehend' how potential futures might feel if they come to be (Andrews 2018). As life moves forward, there always being something new to be 'prehended'.

The vignette above describes how skill is required and employed constantly in professional practice and clinical environments. This is because something is always imminent for which it is needed, and/or to prevent negative situations developing (from becoming imminent). In the ever changing environments of health care, imminence is an everyday part of the job. In terms of feeling and experience, fear of known or unknown exists for learners in a real physical sense. One that is often disturbing but also critical to skills development.

Encounter

Robyn was recently introduced to the new open plan nursing work station. It's a hub where things turn up (such as provisions, new information or directives), people visit and meet (doctors, new patients, families). Sometimes these are collaborative encounters, other times they can be or feel oppositional, and often they present challenges of various forms and scales. Regardless however they bring about change often rapidly. After years in practice Robyn is now an mentor, a job she does at the station by showing mentees how multiple encounters are managed. She is often heard saying "as Mr Smyth to go into room 215" "now call Dr Brooking". "Dot worry about the new rota now, it can wait until later" "Now we need to get those test results".

A scientific understanding of encounter stresses moments when object become proximal enough or physically collide, resulting in changes occurring to them, to their trajectories or beyond to other objects. Of particular interest are the resulting forms and forces, for example, the way in which energy is expelled (through, for example, heat, light, sound, movement). Not too far from this understanding is what with regard to the social world Deleuze—and scholars of NRT—specifically conceptualise as 'events'—meetings of human bodies and objects moving on specific planes with specific trajectories. Events being physical, pre-personal, intensive singularities and important spatial, temporal and processual moments (Duff 2014). In some cases, encounters might be predicable (imminent), particularly if the bodies and objects involved have been formerly rerolled and motivated in known directions by existing structures (such as policies). In other cases, however, encounters might be seemingly random, the bodies and objects involved being motivated more by their own qualities, purposes or impulses, setting themselves on collision courses. The outcomes of encounters are important because they provide a moment/stage for the transformation and reinvention of bodies and objects in relations of becoming (Duff 2014). Indeed, new realities are produced by encounters; new movements, directions and configurations. Each encounter in life is part of a greater web of events (i.e. each is made by previous encounters and influences future encounters) (Duff 2014). Indeed, for Deleuze, the existence of constant and unique encounters (events) means that life is made through 'discontinuous becoming'—a complex and lumpy process rather than a clear and linear one, involving progress and retreat, changes in direction, increases and decreases in capacity with no final mature state (Duff 2014).

These lines of thinking are carried forward more extensively into considerations of social contexts by Dawney (2013) who considers some encounters as 'interruptions'. She argues that interruptions—which are momentary and, almost always to some extent, physical—interrupt the flow of affective less-than-fully conscious experience, as humans suddenly become more consciously aware of their actions and surroundings. Although humans might not always want to interrupt the flow of a situation, and doing this might not always have positive outcomes, interruptions do nonetheless provide humans with an opportunity to consciously take stock, to focus on what went previously and what might occur next—to interplay the affective, cognitive and subjective (Dawney 2013).

The vignette shows how encounters happen in clinical environments between bodies and other bodies and bodies and objects (often that need to be used and mastered). They might be as obvious as a sudden emergency or as everyday as an interaction with a new drug or family member, but all are important as steps in the creeping development of a skills set. This is because skill is acquired in the process and moment of encounter, when something is experienced and dealt with successfully or unsuccessfully. Meanwhile, smaller, seemingly irrelevant encounters act as interruptions, disrupting good and bad flows and creating new ones but always allowing an opportunity to briefly stop and think.

Stillness

Jim is a medical student who, like all medical students, attends lectures and seminars and learns across campus. Although campus can be a lively place in the evenings, during the day it is quiet and relaxed, yet with a serous aire; the perfect setting to learn the theories and facts that Jim is so interested in. Sometimes Jim plays sounds in his headphones which helps him focus. They have to be ambient however like rain, waves, thunder or slow electronic music; something that will not distract or become a focal point in itself. At the weekends Jim often visits a local spa/retreat. There he might relax and think about the things he's learned. He might however just chill and unwind, think of nothing apart what is happening to his body.

According to physics, for an object to be still/at rest, it must lack speed of any perceptible measurement. Newton's First Law of Motion states that an object will remain at rest unless acted upon by an external force. In practice, nothing in life is ever completely still—at absolute rest. It will vibrate, move with subtle atmospheric or environmental effects, its electrons and atoms always move; and ultimately, it is situated on the earth, which is rotating around the sun. As scholars of NRT note, stillness is not just a mathematically defined state of an object; it is also a physical state of the human body and mental state of the human mind, experienced as a sensation (Bissell 2011; Bissell and Fuller 2011). Stillness is the counterpoint to speed, rhythm, vitality, infectiousness, imminence, and encounter—the reference and contrast to these. It lacks the involvements and distractions of these facets of whole onflow. Yet, stillness is not universal. It can be a moment of quiet yet focused reflection and contemplation (on the past or something new), or in contrast a moment of mindfulllness—of freeing oneself from mental preoccupations and ruminations. Stillness then, as an empty opportunity, can be what a human wants it to be.

The final fictional vignette sheds light on how, in the acquisition of skill, stillness is important. Often found in lecture theatres, libraries and study desks where one acquires foundational and theoretical knowledge from talks, books and papers; stillness provides an opportunity to think deeply about complex ideas and form one's own opinions. In other words, these spaces are important, not because of the actual information arising or circulated in them, but because of how they enable the consumption and internalisation of that information. Elsewhere, such as at home or in natural settings, stillness provides the opportunity to break, relax, recover and restore. Both of these experiences of stillness are important to a learner throughout their period of learning, and to their eventual success.

Conclusions

The book has reviewed the long history of geographical thinking on practice (Chap. 1), the motivations, concepts and theories behind contemporary geographical research on practice (Chap. 2) and has shown some diverse empirics demonstrating geographies of professional practice in action at different scales (Chaps. 3–5). In this the final chapter, to showcase further integration in a future research agenda, a connection has been made to an established concept, field and debate in mainstream health professional research—that of skills (Chap. 2 doing the same for PCC). A new ontological understanding of skill is presented—framed as skilling space—that shows how skill both evolves processually in space and time and works in the overall production of spaces of healthcare. The idea of skilling space is not meant, however, as a new or replacement model to Benner's or others. It is more an explanation that showcases a number of important aspects that future research might consider.

We obviously end on a very specific note. For that, however, we do not apologise because this is exactly how most past geographical research on professional practice was framed, and how most future geographical research on professional practice will be. Geography, like any other social science discipline, brings a broad perspective and a way of looking at and unpacking the world of health care. But it is a perspective that has and can look at and contribute to very specific debates and bodies of knowledge.

The future agenda for geographical research on professional practice is wide-ranging. Empirically, scholars must continue to engage with many of the social, technological and service trends that foreshadowed and demanded the establishment of geographical research on practice in the first place—particularly as these trends continue to develop and mutate in the twenty-first century—as well as engage with new ones as they emerge. Conceptually, scholars must continue to consider space and place at different scales and in different ways through a full-range of theoretical lenses, in addition to adopting and contributing to new lenses as they themselves emerge. But, as shown in the final chapters, they might integrate geography further to make sure that the discipline fulfills its full potential, both making sense to new generations of students and professional researchers and playing an even greater role in creating knowledge/evidence that informs practice.

References

Adler, P., & Adler, P. A. (1978). The role of momentum in sport. *Journal of Contemporary Ethnography, 7*(2), 153–175.

Anderson, B. (2009). Affective atmospheres. *Emotion, Space and Society, 2*(2), 77–81.

Andrews, G. J. (2018). *Non-representational theory & health: The health in life in space-time revealing*. Milton Park: Routledge.

Andrews, G. J. (2020). Health geographies III: More-than-representational pushes and expressions. *Progress in Human Geography, 44*(5), 991–1003.

Astor, A., Akhtar, T., Matallana, M. A., Muthuswamy, V., Olowu, F. A., Tallo, V., et al. (2005). Physician migration: Views from professionals in Colombia, Nigeria, India, Pakistan and the Philippines. *Social Science & Medicine, 61*(12), 2492–2500.

Banning, M. (2008). A review of clinical decision making: Models and current research. *Journal of Clinical Nursing, 17*(2), 187–195.

Benner, P. (1982). From novice to expert. *AJN The American Journal of Nursing, 82*(3), 402–407.

Benner, P. (2004). Using the Dreyfus model of skill acquisition to describe and interpret skill acquisition and clinical judgment in nursing practice and education. *Bulletin of Science, Technology & Society, 24*(3), 188–199.

Benner, P. E. (1984). *From novice to expert: Excellence and power in clinical nursing practice*. Boston: Addison-Wesley.

Bennett, J. (2010). *Vibrant matter: A political ecology of things*. Durham: Duke University Press.

Bergson, H. (1911). *Matter and memory*. Crows Nest: Unwin.

Bissell, D. (2011). Thinking habits for uncertain subjects: Movement, stillness, susceptibility. *Environment and Planning A, 43*, 2649–2665.

Bissell, D., & Fuller, G. (Eds.). (2011). *Stillness in a mobile world*. Milton Park: Routledge.

Bourgeois, M. S., Dijkstra, K., Burgio, L. D., & Allen, R. S. (2004). Communication skills training for nursing aides of residents with dementia: The impact of measuring performance. *Clinical Gerontologist, 27*(1-2), 119–138.

Bowles, N., Mackintosh, C., & Torn, A. (2001). Nurses' communication skills: An evaluation of the impact of solution-focused communication training. *Journal of Advanced Nursing, 36*(3), 347–354.

Brocklehurst, P., & Tickle, M. (2011). The policy context for skill mix in the National Health Service in the United Kingdom. *British Dental Journal, 211*(6), 265.

Buchan, J. (2006). The impact of global nursing migration on health services delivery. *Policy, Politics, & Nursing Practice, 7*(3_Suppl), 16S–25S.

Buchan, J., & Dal Poz, M. R. (2002). Skill mix in the health care workforce: Reviewing the evidence. *Bulletin of the World Health Organization, 80*, 575–580.

Castle, A. (2006). Assessment of the critical thinking skills of student radiographers. *Radiography, 12*(2), 88–95.

Chang, C. C., & Lin, L. C. (2005). Effects of a feeding skills training programme on nursing assistants and dementia patients. *Journal of Clinical Nursing, 14*(10), 1185–1192.

College of Nurses of Ontario (2014) Entry to practice competencies https://www.cno.org/globalassets/docs/reg/41037_entrytopracitic_final.pdf

Courtney, K. L., Alexander, G. L., & Demiris, G. (2008). Information technology from novice to expert: Implementation implications. *Journal of Nursing Management, 16*(6), 692–699.

Crossan, F., & Ferguson, D. (2005). Exploring nursing skill mix: A review. *Journal of Nursing Management, 13*(4), 356–362.

Dalton, J. A. (1989). Nurses' perceptions of their pain assessment skills, pain management practices, and attitudes toward pain. *Oncology Nursing Forum, 16*(2), 225–231.

Darbyshire, P. (1994). Skilled expert practice: Is it 'all in the mind'? A response to English's critique of Benner's novice to expert model. *Journal of Advanced Nursing, 19*(4), 755–761.

Davies, N., & Gould, D. (2000). Updating cardiopulmonary resuscitation skills: A study to examine the efficacy of self-instruction on nurses' competence. *Journal of Clinical Nursing, 9*(3), 400–410.

Dawney, L. (2013). The interruption: Investigating subjectivation and affect. *Environment and Planning D: Society and Space, 31*(4), 628–644.

Delaney, K. R., & Johnson, M. E. (2006). Keeping the unit safe: Mapping psychiatric nursing skills. *Journal of the American Psychiatric Nurses Association, 12*(4), 198–207.

Deleuze, G., & Guattari, F. (1988). *A thousand plateaus: Capitalism and schizophrenia*. London: Bloomsbury Publishing.

Deleuze, G., & Parnet, C. (2006). *Dialogues II*. London: Continuum.

Dolan, G. (2003). Assessing student nurse clinical competency: Will we ever get it right? *Journal of Clinical Nursing, 12*(1), 132–141.

Dornan, T., Lee, C., & Stopford, A. (2001). Skills base: A web-based electronic learning portfolio for clinical skills. *Academic Medicine, 76*(5), 542–543.

Dreyfus, S. E., & Dreyfus, H. L. (1980). *A five-stage model of the mental activities involved in directed skill acquisition* (No. ORC-80-2). California Univ Berkeley Operations Research Center.

Dubois, C. A., & Singh, D. (2009). From staff-mix to skill-mix and beyond: Towards a systemic approach to health workforce management. *Human Resources for Health, 7*(1), 87–92.

Duff, C. (2014). *Assemblages of health: Deleuze's empiricism and the ethology of life*. New York: Springer.

Duffield, C. M., Forbes-Madden, J. L., Fallon, A., Roche, M. A., Wise, W., & Merrick, E. T. (2005). Nursing skill mix and nursing time: The roles of registered nurses and clinical nurse specialists. *Australian Journal of Advanced Nursing, 23*(2), 14–21.

English, I. (1993). Intuition as a function of the expert nurse: A critique of Benner's novice to expert model. *Journal of Advanced Nursing, 18*(3), 387–393.

Gaies, M. G., Landrigan, C. P., Hafler, J. P., & Sandora, T. J. (2007). Assessing procedural skills training in pediatric residency programs. *Pediatrics, 120*(4), 715–722.

Gardner, L. (2012). From novice to expert: Benner's legacy for nurse education. *Nurse Education Today, 32*(4), 339–340.

Gatley, E. P. (1992). From novice to expert: The use of intuitive knowledge as a basis for district nurse education. *Nurse Education Today, 12*(2), 81–87.

Gauntlett, R., & Laws, D. (2008). Communication skills in critical care. *Continuing Education in Anaesthesia, Critical Care & Pain, 8*(4), 121–124.

Gijbels, M. (1995). Mental health nursing skills in an acute admission environment: Perceptions of mental health nurses and other mental health professionals. *Journal of Advanced Nursing, 21*(3), 460–465.

Glen, S., & Clark, A. (1999). Nurse education: A skill mix for the future. *Nurse Education Today, 19*(1), 12–19.

Glover, M. L., & Sussmane, J. B. (2002). Assessing pediatrics residents' mathematical skills for prescribing medication: A need for improved training. *Academic Medicine, 77*(10), 1007–1010.

Gobet, F., & Chassy, P. (2008). Towards an alternative to Benner's theory of expert intuition in nursing: A discussion paper. *International Journal of Nursing Studies, 45*(1), 129–139.

Godsell, G. (2004). A nurse-surgical post cuts waiting times and extends nurses' skills base. *Professional Nurse, 19*(8), 453–455.

Haag-Heitman, B. (1999). *Clinical practice development using novice to expert theory*. Burlington: Jones & Bartlett Learning.

Hall, E., & Wilton, R. (2017). Towards a relational geography of disability. *Progress in Human Geography, 41*(6), 727–744.

Han, H., & Hyun, S. S. (2015). Customer retention in the medical tourism industry: Impact of quality, satisfaction, trust, and price reasonableness. *Tourism Management, 46*, 20–29.

Hardy, M., & Snaith, B. (2006). Role extension and role advancement—Is there a difference? A discussion paper. *Radiography, 12*(4), 327–331.

Harper, P. R., Powell, N. H., & Williams, J. E. (2010). Modelling the size and skill-mix of hospital nursing teams. *Journal of the Operational Research Society, 61*(5), 768–779.

Harris, M. A., Pittiglio, L., Newton, S. E., & Moore, G. (2014). Using simulation to improve the medication administration skills of undergraduate nursing students. *Nursing Education Perspectives, 35*(1), 26–29.

Hignett, S. (2003). Intervention strategies to reduce musculoskeletal injuries associated with handling patients: A systematic review. *Occupational and Environmental Medicine, 60*(9), e6–e6.

Hird, V. (1995). Nursing competencies: The artistry of nursing. Retrieved from http://www.ciap.health.nsw.gov.au/hospolic/stvincents/1995/a05.html

Hobbs, S. D. (2002). Measuring nurses' computer competency: An analysis of published instruments. *CIN: Computers, Informatics, Nursing, 20*(2), 63–73.

Ingold, T. (2000). *The perception of the environment: Essays on livelihood, dwelling and skill.* Milton Park: Routledge.

Ingold, T. (2013). *Making: Anthropology, archaeology, art and architecture.* Milton Park: Routledge.

Jeffries, P. R. (2001). Computer versus lecture: A comparison of two methods of teaching oral medication administration in a nursing skills laboratory. *Journal of Nursing Education, 40*(7), 323–329.

Jeffries, P. R., Rew, S., & Cramer, J. M. (2002). A comparison of student-centered versus traditional methods of teaching basic nursing skills in a learning laboratory. *Nursing Education Perspectives, 23*(1), 14–19.

Jenkins-Clarke, S., Carr-Hill, R., & Dixon, P. (1998). Teams and seams: skill mix in primary care. Journal of Advanced Nursing, 28(5), 1120–1126.

Jenkins, J., & Calzone, K. A. (2007). Establishing the essential nursing competencies for genetics and genomics. *Journal of Nursing Scholarship, 39*(1), 10–16.

Jennett, P. A., Watson, M. M., & Watanabe, M. (2000). The potential effects of telehealth on the Canadian health workforce: Where is the evidence? *CyberPsychology & Behavior, 3*(6), 917–923.

King's College. (2009). *Nursing competencies.* London: King's College.

Kingma, M. (2001). Nursing migration: Global treasure hunt or disaster-in-the-making? *Nursing Inquiry, 8*(4), 205–212.

Kneafsey, R., & Haigh, C. (2007). Learning safe patient handling skills: Student nurse experiences of university and practice based education. *Nurse Education Today, 27*(8), 832–839.

Latter, S., Rycroft-Malone, J., Yerrell, P., & Shaw, D. (2001). Nurses' educational preparation for a medication education role: Findings from a national survey. *Nurse Education Today, 21*(2), 143–154.

Latter, S., Yerrell, P., Rycroft-Malone, J., & Shaw, D. (2000). Nursing, medication education and the new policy agenda: The evidence base. *International Journal of Nursing Studies, 37*(6), 469–479.

Lauri, S., & Salanterä, S. (1998). Decision-making models in different fields of nursing. *Research in Nursing & Health, 21*(5), 443–452.

Layman Young, J., Horton, F. M., & Davidhizar, R. (2006). Nursing attitudes and beliefs in pain assessment and management. *Journal of Advanced Nursing, 53*(4), 412–421.

Lefebvre, H. (1992). *Éléments de rythmanalyse.* Paris: Éditions Syllepse.

Lehmann, U., & Makhanya, N. (2005). Building the skills base to implement the district health system: Human resources. *South African Health Review, 2005*(1), 136–145.

Lima, S., Newall, F., Kinney, S., Jordan, H. L., & Hamilton, B. (2014). How competent are they? Graduate nurses self-assessment of competence at the start of their careers. *Collegian, 21*(4), 353–358.

Lockwood, P. (2016). An economic evaluation of introducing a skills mix approach to CT head reporting in clinical practice. *Radiography, 22*(2), 124–130.

Madden, C. (2006). Undergraduate nursing students' acquisition and retention of CPR knowledge and skills. *Nurse Education Today, 26*(3), 218–227.

Manley, K., & Garbett, R. (2000). Paying Peter and Paul: Reconciling concepts of expertise with competency for a clinical career structure. *Journal of Clinical Nursing, 9*(3), 347–359.
Mason, T., Coyle, D., & Lovell, A. (2008). Forensic psychiatric nursing: Skills and competencies: II. Clinical aspects. *Journal of Psychiatric and Mental Health Nursing, 15*(2), 131–113.
Mauk, K. (Ed.). (2010). *Gerontological nursing: Competencies for care*. Burlington: Jones & Bartlett Learning.
McMullan, M., Endacott, R., Gray, M. A., Jasper, M., Miller, C. M., Scholes, J., et al. (2003). Portfolios and assessment of competence: A review of the literature. *Journal of Advanced Nursing, 41*(3), 283–294.
Mid Staffordshire NHS Foundation Trust. (2013). *Report of the Mid Staffordshire NHS Foundation Trust Public Inquiry: Executive summary*. Retrieved from http://www.official-documents.gov.uk/document/hc1213/hc09/0947/0947.pdf
Miller, M., & Scully, C. (2015). *Mosby's textbook of dental nursing*. New York: Elsevier Health Sciences.
Mossey, P. A., Newton, J. P., & Stirrups, D. R. (1997). Defining, conferring and assessing the skills of the dentist. *British Dental Journal, 182*(4), 123.
Odell, M., Victor, C., & Oliver, D. (2009). Nurses' role in detecting deterioration in ward patients: Systematic literature review. *Journal of Advanced Nursing, 65*(10), 1992–2006.
Oliver, T. K., Butzin, D. W., Guerin, R. O., & Brownlee, R. C. (1991). Technical skills required in general pediatric practice. *Pediatrics, 88*(4), 670–673.
Paley, J. (1996). Intuition and expertise: Comments on the Benner debate. *Journal of Advanced Nursing, 23*(4), 665–671.
Patchett, M. (2016). The taxidermist's apprentice: Stitching together the past and present of a craft practice. *Cultural Geographies, 23*(3), 401–419.
Patchett, M., & Mann, J. (2018). Five advantages of skill. *Cultural Geographies, 25*(1), 23–29.
Perry, A. G., Potter, P. A., & Ostendorf, W. (2013). *Clinical nursing skills and techniques*. New York: Elsevier Health Sciences.
Philo, C. (2007). A vitally human medical geography? Introducing Georges Canguilhem to geographers. *New Zealand Geographer, 63*(2), 82–96.
Pile, S. (2010). Emotions and affect in recent human geography. *Transactions of the Institute of British Geographers, 35*(1), 5–20.
Polivka, B. J., Stanley, S. A., Gordon, D., Taulbee, K., Kieffer, G., & McCorkle, S. M. (2008). Public health nursing competencies for public health surge events. *Public Health Nursing, 25*(2), 159–165.
Pratt, S., & Adams, C. (2003). How to create a degree course in radiography: A recipe. *Radiography, 9*(4), 317–322.
Profetto-McGrath, J. (2003). The relationship of critical thinking skills and critical thinking dispositions of baccalaureate nursing students. *Journal of Advanced Nursing, 43*(6), 569–577.
Reader, T., Flin, R., Lauche, K., & Cuthbertson, B. H. (2006). Non-technical skills in the intensive care unit. *British Journal of Anaesthesia, 96*(5), 551–559.
Reader, T. W., Flin, R., & Cuthbertson, B. H. (2007). Communication skills and error in the intensive care unit. *Current Opinion in Critical Care, 13*(6), 732–736.
Robbins, J. W., Hilton, T. J., Schwartz, R. S., & dos Santos Jr, J. (2006). Fundamentals of operative dentistry: a contemporary approach. J. B. Summitt (Ed.). Quintessence Pub.
Shanahan, M. C. (2007). Information literacy skills of undergraduate medical radiation students. *Radiography, 13*(3), 187–196.
Shin, K. R. (1998). Critical thinking ability and clinical decision-making skills among senior nursing students in associate and baccalaureate programmes in Korea. *Journal of Advanced Nursing, 27*(2), 414–418.
Smith, S. F., Duell, D., Martin, B. C., Aebersold, M. L., & Gonzalez, L. (2008). *Clinical nursing skills: Basic to advanced skills*. Upper Saddle River: Pearson Prentice Hall.
Spielman, A. I., Fulmer, T., Eisenberg, E. S., & Alfano, M. C. (2005). Dentistry, nursing, and medicine: A comparison of core competencies. *Journal of Dental Education, 69*(11), 1257–1271.

Spilsbury, K., & Meyer, J. (2001). Defining the nursing contribution to patient outcome: Lessons from a review of the literature examining nursing outcomes, skill mix and changing roles. *Journal of Clinical Nursing, 10*(1), 3–14.

Standing, M. (2007). Clinical decision-making skills on the developmental journey from student to Registered Nurse: A longitudinal inquiry. *Journal of Advanced Nursing, 60*(3), 257–269.

Supic, Z. T., Bjegovic, V., Marinkovic, J., Milicevic, M. S., & Vasic, V. (2010). Hospital management training and improvement in managerial skills: Serbian experience. *Health Policy, 96*(1), 80–89.

Thompson, C. (1999). A conceptual treadmill: The need for 'middle ground' in clinical decision making theory in nursing. *Journal of Advanced Nursing, 30*(5), 1222–1229.

Thrift, N. (2008). *Non-representational theory: Space, politics, affect.* London: Routledge.

Turner, L. (2010). "Medical tourism" and the global marketplace in health services: US patients, international hospitals, and the search for affordable health care. *International Journal of Health Services, 40*(3), 443–467.

Umiker, W. O., & Umiker, W. (2005). *Management skills for the new health care supervisor.* Sudbury: Jones & Bartlett Learning.

Van Leeuwen, R., & Cusveller, B. (2004). Nursing competencies for spiritual care. *Journal of Advanced Nursing, 48*(3), 234–246.

Watson, R., Stimpson, A., Topping, A., & Porock, D. (2002). Clinical competence assessment in nursing: A systematic review of the literature. *Journal of Advanced Nursing, 39*(5), 421–431.

Wilgis, M., & McConnell, J. (2008). Concept mapping: An educational strategy to improve graduate nurses' critical thinking skills during a hospital orientation program. *The Journal of Continuing Education in Nursing, 39*(3), 119–126.

Wilson, C. B. (2001). Safer handling practice for nurses: A review of the literature. *British Journal of Nursing, 10*(2), 108–114.

Zhang, Z. X., Luk, W., Arthur, D., & Wong, T. (2001). Nursing competencies: Personal characteristics contributing to effective nursing performance. *Journal of Advanced Nursing, 33*(4), 467–474.

Index

G. J. Andrews et al., *Place and Professional Practice*, Global Perspectives on Health Geography, https://doi.org/10.1007/978-3-030-64179-5

E

F

G